KAIST Research Series

The KAIST Research Series has been established to enable the wide dissemination of KAIST's outstanding research achievements among researchers at an international level. It provides a framework for KAIST researchers to publish monographs synthesizing the key results of their extended research programmes and setting these in a global context. Volumes published in the KAIST Research Series include single- and joint-authored monographs written by KAIST researchers, and edited volumes arranged by KAIST academics but including an international authorship. The KAIST Research Series also provides a forum for undergraduate-level textbooks and graduate-level textbooks based around lecture courses given at KAIST. It is coordinated by a team of Series Editors, representing most academic departments within KAIST.

About KAIST KAIST was established by the government in 1971 as the nation's first graduate school specializing in science and engineering education and research. By 2006, KAIST has produced over 30,000 graduates. A large number of KAIST graduates have become researchers, venture entrepreneurs and technological bureaucrats, contributing to Korea's economic growth and development of science and technology. KAIST ranked seventh in Asia in a ranking compiled jointly by the Chosun Ilbo, a major Korean daily, and global university evaluation institute QS of Britain, and was included in the "2012 Top 100 Global Innovative Organizations" published by Thomson Reuters, the world's leading media and financial-data firm. KAIST is committed to becoming one of the world's leading universities focused on science and technology with a vision to create knowledge for human society.

About KAIST Press KAIST Press is a university press established to disseminate KAIST's outstanding research achievements and popularize science & technology. KAIST Press has published research papers, reports and teaching materials, as well as books aimed at public understanding of science and technology. It has recently expanded its operations into organizing a series of guest lectures designed to promote public understanding of science and technology.

Hyungjun Cho

The AI as Social Actor

Design Speculations for Reimagining Human-Thing Relationships

Springer

Hyungjun Cho
School of Art + Art History
University of Florida
Gainesville, FL, USA

ISSN 2214-2541 ISSN 2214-255X (electronic)
KAIST Research Series
ISBN 978-981-92-0536-3 ISBN 978-981-92-0537-0 (eBook)
https://doi.org/10.1007/978-981-92-0537-0

This Springer imprint is published by the registered company Springer Nature Singapore Pte Ltd.
The registered company address is: 152 Beach Road, #21-01/04 Gateway East, Singapore 189721, Singapore

Preface

This book is a reworking of my doctoral dissertation. Since less than a year has passed since I received my PhD, it would not be inaccurate to say that this book traces my journey as a researcher. I never expected to write a book so early in my career, and I am deeply grateful for the opportunity to do so.

I completed my undergraduate, master's, and doctoral studies in the Department of Industrial Design at KAIST. As the name suggests, the institution is strongly oriented toward science and engineering. Many of my peers who pursued doctoral studies worked on topics such as developing new machine learning algorithms, mathematically modeling cellular differentiation, or improving semiconductor manufacturing processes. When I met them, and the conversation turned to research, I often found myself explaining that I was studying how people perceive robots that post on social media. While this description was not incorrect, it was hardly flattering. I remain thankful that my friends treated this work with respect rather than dismissal.

Looking back, I am not entirely sure what kind of research I originally wanted to pursue. Nevertheless, from the beginning, my interests have been remarkably consistent. I was inspired by films such as *Big Hero*, *Iron Man*, *Her*, and *Chappie*, and I wanted to be involved in the design of such beings. Translating this interest into research language, however, proved unexpectedly difficult. At various moments, I framed my work as expressive interaction for product attachment, attempted to develop a design methodology based on the concept of tsukumogami, or immersed myself in philosophy, including ontology and existentialism, through thinkers such as Heidegger, Sartre, and Wittgenstein. None of these framings traveled well. I was repeatedly asked why this research was necessary or what gap it addressed. These questions came from colleagues and from peer review alike. As a result, I developed a strong resistance to viewing design and research solely through a problem-solving frame, even though I recognize that this frame can be valuable in other contexts.

It was only within the past one or two years that I encountered speculative design and learned how to articulate the contributions of my work to the academic community. The idea that design does not need to solve problems, but can instead generate discourse and provoke discussion, was immediately compelling. I feel a sense of gratitude toward Dunne and Raby, despite never having met them in person.

Even now, I am still learning how to understand my own research topics and approaches. I continue to be surprised by how differently the same work can be received depending on how it is framed and communicated. Recently, I have become deeply engaged in distinguishing speculative design from future prediction. At the thematic level, my work has shifted from being described as human-AI interaction to a focus on relationships, and more recently, to engaging with concepts from Haraway and Barad. In this process, the research I had already conducted has begun to function as speculative design for me personally, prompting alternative ways of thinking rather than confirming established positions. Perhaps this suggests that the design itself was effective.

I am still growing as a researcher and am only beginning to find my place. Even so, I believe I have reached a point where I can present the design research I conducted during my doctoral studies in a meaningful way and articulate the insights that can be drawn from it. The person writing this book is no longer the same person who wrote the dissertation. I write now with greater confidence in the depth of my thinking, particularly in the third part of the book, which introduces an entirely new line of discussion.

What I aim to do in this book is neither to establish a new theory nor to propose a novel perspective. Instead, I introduce three design research cases and demonstrate that interactions between humans and things, particularly those embedded with AI, are no longer a matter of simple tools. They are complex social relationships. I discuss why and how these relationships become social, and what kinds of orientations might guide those who design such things. I do not believe these discussions offer definitive answers.

What I hope, instead, is that my designs will be engaging to readers. I hope they invite readers to interpret the designs on their own, to draw inspiration from them, and to arrive at ways of thinking that differ from their own. Compared to the artifacts produced by many design researchers I admire, my work is relatively direct. I have not yet to learn how to conceal dense philosophy within abstract forms and interactions. For that reason, I believe these designs remain accessible and capable of inviting many people into a shared line of questioning. The question I pose in this book is as direct as the artifacts themselves. In the near future, what will the relationship between humans and things be like?

Gainesville, FL, USA
December 2025

Hyungjun Cho

Competing Interests The author has no competing interests to declare that are relevant to the content of this manuscript.

Contents

Acronyms

AI	Artificial Intelligence
HCI	Human-Computer Interaction
HRI	Human-Robot Interaction
LLM	Large Language Model
RtD	Research through Design
STS	Science and Technology Studies

Part I
Rethinking Relations with AI

This part lays the conceptual and methodological groundwork for this book. As AI endows linguistic capabilities to things, they gradually begin to be experienced as relational entities beyond mere tools. This shift is the core concern and starting point that runs through this book. Chapter 1 examines, from the perspective of a researcher situated at the intersection of human-computer interaction (HCI) and design, how people form social and emotional relationships with AI even amid the inertia of still understanding technology as a tool. Chapter 2 then explains why this book turns to research through design and speculative design not as stylistic choices, but as methods for inquiry that reveal the possibilities of relationships.

Chapter 1
When AI Gives Things a Voice

Abstract This chapter examines a dramatic shift that underpins my research. While we continue to describe technology as a tool, we increasingly experience it as an entity with which we socially interact and form relationships. Large language models (LLMs), in particular, use language to make things feel like entities that perform social roles, such as empathy, comfort, advice, and authority, rather than merely functional tools. As a result, personification is no longer an exceptional reaction but has become almost automatic. I trace the origins of this shift by examining the long-standing inertia in design and usability discourse that prioritizes utility, alongside accumulated research on personification that shows how people project social norms and personal intentions onto technology even within this framework. I then explore how personification expands from function to role in robots and automated systems, and how language-capable AI amplifies this tendency to an unprecedented degree. Against this backdrop, the chapter argues for a shift in focus away from evaluating how well AI mimics humans and toward examining the expectations, emotions, and relationships of responsibility and dependence that people form through these technologies. This shift provides the conceptual starting point that runs through the design research case studies presented in the subsequent chapters.

1.1 Anthropomorphism in Use

Over long periods, humans have created a wide range of things. This activity has been understood as a means to meet human needs, increase convenience in work, adapt to surrounding environments, and, at times, express creativity [6, 37]. Some have traced the history of such design activity back to the very beginnings of civilization [9, 28]. Still, it is generally accepted that what we now call design, as it was named and institutionalized as a professional domain, took shape between the mid-eighteenth and nineteenth centuries during the Industrial Revolution.

As craft traditions, in which individuals made singular items for specific people, became incorporated into systems of mass production for an anonymous public, design gradually developed as a capability for making better things. Today, the scope of design encompasses graphics and products, extending to large-scale applications

H. Cho, *The AI as Social Actor*, KAIST Research Series,
https://doi.org/10.1007/978-981-92-0537-0_1

such as architecture, and addresses user interfaces and interactions for digital products. It also reaches into intangible domains, including user experience, customer service, and even policy. In this sense, the range of targets for improvement, in other words, design itself, has continued to expand.

In short, all artificial artifacts created by someone can be considered things in a broad sense, and the act of creating them can be understood as design, regardless of the intended target. This broad definition resonates with Wakkary's discussion [52]. He defines a thing as a nonhuman entity created by an actor, whether human or nonhuman, such as algorithms or autonomous systems. In this book, I likewise begin with a broad understanding of things, which includes not only physical artifacts made by humans but also technologies such as software systems.

As design has evolved into specialized professional domains, the field of design discourse has also developed as an academic discipline. Today, this has led to a set of widely shared principles on how we should design things. In industrial design and user experience, in particular, a dominant view holds that things should perform their intended functions well and should therefore be designed to be intuitive and easy to use [22, 35]. Maintained for decades as nearly absolute standards, these principles have reinforced a production-centered view of design and contributed to contemporary values in which it feels natural to see things as tools.

For example, when we encounter a particular thing, we tend to focus on how we can use it and evaluate it based on its usefulness. This perspective can be well illustrated through Heidegger's analysis of the hammer [20], although his original aim was ontologically far more profound. Through his analysis, Heidegger suggested that the primary way humans relate to things lies in their use or usefulness, and that a hammer reveals its mode of being as a tool when it is grasped, receding transparently into the background. Later, this tool-centered perspective was criticized by philosophers who sought to emphasize the ontological status of things themselves, including those associated with object-oriented ontology, such as Harman [19]. Even so, it is difficult to deny that most people today still recognize and describe things primarily through a tool user frame.

Yet, just as expectations and reality do not always align, and just as human instincts are not always rational, the ways humans relate to things cannot be fully reduced to function-centered use. A representative deviation from this pattern is anthropomorphism. Nass and his colleagues proposed early on that people respond to media as if they were interacting with real people or physical environments [33]. Their experiments showed that even when using technology, people expect it to follow social norms, such as politeness and distinctions between the self and others.

This human tendency to anthropomorphize technology becomes more pronounced as technology grows more advanced. Rule-based automation systems, for instance, are often understood as entities with intentions. Building on this tendency, Rozendaal proposed the concept of Objects with Intent, in which everyday things display intelligence and agency, acting as collaborative partners [42]. Examples include a lamp that dims itself to help a user fall asleep, or a jacket that provides biofeedback through its material and tactile qualities to help the wearer calm down in response to stress levels. Many other studies have repeatedly shown that people tend to attribute desires, thoughts, and emotions to technological systems, assume that they can perceive situations, and believe that they possess intelligence for autonomous decision-making [10, 13, 16, 36, 44].

Furthermore, in the field of human-robot interaction (HRI), anthropomorphism has already become a primary design paradigm. Through facial expressions [4, 40], gestures [12, 21], gaze behaviors [27, 31], and proxemics [32], practitioners design robots that mimic human behavior. Users, in turn, expect something humanlike from robots, while robots increasingly take on human roles even in functional terms. There is no clear causal order here. Still, the tendency to anthropomorphize robots is not unrelated to the expansion of robotic functions into domains traditionally associated with human roles. As a result, robots are expected to perform social roles, such as butler [5], guide [8], or caregiver [49].

These advanced technologies pose substantial challenges for practitioners and researchers who design the things that embed them. The problem lies in the significant gap between the actual operating mechanisms of these devices and systems and the intuitive inner states or responses that people imagine through anthropomorphism. For instance, if a robot turns its head toward us, smiles, and waves, we may instinctively feel that it is welcoming us. In reality, however, it is far more likely that the behavior is a preprogrammed response triggered by motion detection, and the likelihood that the robot has an intention in the sense of welcoming us is exceptionally low.

Because of this gap, human-computer interaction (HCI) and HRI research have long emphasized the importance of design strategies that carefully calibrate the level of anthropomorphism to prevent excessive expectations and misunderstandings from undermining usability or user experience. This line of research has accumulated considerable knowledge and practice, and it remains an ongoing struggle. Anthropomorphism is closely tied to human nature, and from a designer's perspective, it is an effective strategy for fostering trust in systems [34], increasing intimacy [39, 47], and enhancing usability [11]. If the degree of anthropomorphism can be finely tuned, complex systems can become more predictable, cognitive burden can be reduced, and overall satisfaction can be improved.

1.2 Being Social by Using Language

In short, anthropomorphizing things is deeply rooted in human nature, and researchers have long explored these psychological responses to technology. Recently, however, AI, and especially talking AI, has intensified this issue to an unprecedented degree.

To clarify, AI is a broad term that encompasses various technologies, including machine learning, deep learning, and generative models. In this book, however, AI refers more narrowly to large language models and interactive systems that incorporate them. When people think of AI today, language model-based technologies are what most often come to mind, and they are also what have produced this level of impact across industry and popular culture. Beyond their familiarity, the reason for this focus is that the book's core theme, AI as a social entity, cannot be discussed without considering the central role of language.

The influence of language on intelligent and social human activity is so significant that it almost feels redundant to emphasize it. Language is the medium through which individuals most directly and precisely express what they think, feel, and want. Some philosophers have even argued the reverse. Wittgenstein famously stated that the limits of my language mean the limits of my world, suggesting that what we can say defines what we can think [54]. In other words, humans do not simply speak because they think. Rather, they can develop thought in particular ways because they can speak.

In a similar vein, psychologists and cognitive scientists explain that the words, grammar, and metaphors we use organize our emotions and provide the frameworks through which we distinguish and understand concepts [2]. Just as being able to name and distinguish between sadness and disappointment leads us to experience them as distinct emotions, language shapes how emotions are perceived. In this sense, language is a vq07key mechanism that externalizes traits we have considered uniquely human, such as thought, will, desire, and emotion. If such linguistic capability is granted to a nonhuman entity, allowing it to have a voice and speak, this suggests that characteristics we have associated with humanity may no longer be exclusive to humans.

There are, of course, persistent warnings against equating AI with humans. Many in academia and industry emphasize that the fact that AI can discuss something does not mean it understands it. They argue that social experience is ultimately an illusion produced by human interpretation, and that AI is merely an algorithm that reproduces form without regard for real meaning or context [3, 30, 38, 48]. Yet the key issue here is not whether AI truly knows or feels. Instead, it is how people actually experience technologies that appear excessively humanlike, and how we should design things that are likely to display strong humanlike features.

This is because when things begin to use human language, their ability to elicit emotional, social, and cognitive responses increases dramatically. As a result, AI induces anthropomorphism not as a strategic choice to understand complex systems, but as an automatic and instinctive reaction that bypasses technical reasoning. A widely known case involves an engineer at a major technology company who claimed that a language model had become conscious and was subsequently dismissed [51]. A 2023 study reported that when participants were asked whether various currently available interactive technologies seemed to possess consciousness, nearly 75 percent responded that GPT-3 had some degree of consciousness. This was higher than responses to the same question about robotic vacuum cleaners, which stood at 50 percent [45].

Increasingly, we expect AI to do things that we once expected only from other humans, such as empathy, comfort, advice, recognition, intimacy, and even romantic feelings. Remarkably, AI is capable of performing some of these human roles to a certain extent. A 2025 industry report identified therapy and companionship as the most common uses of generative AI [55]. On platforms such as Character.ai, numerous AI characters acting as romantic partners, friends, or mentors are already in active use. In parallel, various AI-based mental health chatbots and self-care tools have become established commercial services. Researchers have also begun to explore these possibilities by examining the opportunities and risks of AI as a romantic partner [53], evaluating the effectiveness and limitations of AI conversational agents as healthcare providers [29], and discussing whether AI can replace or support teachers and tutors [24, 46].

From these cases, I would like to emphasize two key points. First, interactions between humans and AI are increasingly resembling complex and subtle interpersonal dynamics. People infer intentions from an AI's tone, response speed, memory, and contextual understanding. They expect responses to be adjusted to their emotional states, and they experience the same interaction differently depending on perceived levels of intimacy or trust. This is no longer purely a mechanical use in which outputs follow commands. It is closer to social interaction shaped by signals and interpretations.

Second, despite this, we still refer to these AIs as tools and claim that we use them. While this habitual language may not immediately lead to visible disasters, it creates a growing misalignment between how people actually relate to AI and the language used to describe that relationship. This misalignment suggests the need to reexamine our existing assumptions, which understand and design things merely as objects of use. Indeed, scholars have continued to challenge the human tool frame by rethinking human-thing interactions from new perspectives [25, 41, 43]. They argue that things should be seen not as passive targets under human control, but as entities with their own temporality and agency. The approach of this book, which explores relationships between humans and AI, follows this line of thought and will be discussed in more detail in the following subchapter.

1.3 Shifting Focus Toward the Relationship

With the emergence of AI technologies, human-thing interaction has been radically reconfigured. Many technologies now penetrate so deeply into our lives that it feels awkward to call them merely tools. As seen above, drawing a clear line by saying we are human and that is not merely a thing is becoming less convincing. In this context, new frameworks of thought that move beyond human-centered perspectives and first attend to how entities are entangled with one another have gained attention.

One of the clearest starting points is Haraway. For a long time, she has focused on cyborgs and companion species as beings that blur the boundary between human and nonhuman. In Haraway's account, the cyborg traverses the boundaries between human and animal, animal and machine, material and immaterial, and has even been described as our ontology and our politics [18]. Although her work does not directly target contemporary AI or LLMs, her descriptions of beings in which distinctions between human and nonhuman become ambiguous resonate strikingly with the present situation, in which AI is experienced not merely as an object of anthropomorphism but as something increasingly difficult to distinguish from humans.

In her discussion of companion species, Haraway describes how humans and dogs live together through interdependent relationships of mutual training and domestication [17]. The key here is not simply what dogs are, but how humans and dogs shape one another within a network of relationships, how they see each other, and what roles they assign. If we substitute dogs with AI in this scene, it becomes easier to see why attention to expected roles, distance, and responsibility between humans and AI matters. Humans are not subjects standing outside technology and controlling it. They are beings who are co-constituted alongside technology.

This turn toward relationships has appeared simultaneously under various names in science and technology studies and philosophy. Actor-network theory, for example, understands society not as a predefined structure or hierarchy, but as a series of networked connections among humans and nonhumans [26]. Even mundane items, such as seatbelts or door closers, are seen as actors that embody particular values and norms, organizing human behavior. Around the same time, post-phenomenology, developed by thinkers such as Ihde and Verbeek, emphasized that our experience of the world is always mediated by technology, calling for attention to the human-technology relationship [23, 50].

Consider voice assistants on smartphones or smart speakers at home. They are not simply tools that receive voice commands. They reshape what we feel is worth saying, what we imagine we can ask, and what kinds of responses count as conversational. In other words, such systems quietly adjust how we use language to interact with the world and redefine what patterns of speaking and listening feel natural. All of these theories overturn the view that treats humans and technologies as independent entities with relationships added afterward. Instead, they propose starting from entanglement, a condition in which humans and technologies are intertwined from the outset.

Barad pushes this entanglement even further through relational ontology [1]. While traditional views assume that preexisting entities enter into relationships, Barad argues that entities do not exist independently in the first place but emerge only through relations. She distinguishes this from interaction by using the term intra action, emphasizing that humans and nonhumans, subjects and objects, are continually reconfigured through mediations such as experimental apparatuses, algorithms, and social norms.

From this perspective, things such as AI-enabled voice agents are better understood not as talking machines that imitate humans or as new entities that have reached human-level intelligence. Instead, they are part of a relational arrangement formed by datasets, models, interfaces, usage rules, and everyday practices. Within this arrangement, people sometimes treat AI as a tool and at other times as an advisor or a friend, adjusting their tone, expectations, and emotional responses accordingly. What kinds of voices feel natural, how much emotion is acceptable to express, and similar judgments are coproduced within these human-AI assemblages.

This relational perspective has also gained traction in HCI and design research. Frauenberger argues that instead of separating users and technologies, design should be approached within living entanglements where people, technologies, norms, and infrastructures are deeply interwoven [15]. Design, in this view, is not about improving specific functions but about intervening in and rearranging existing networks of relations. Even a small device in a room or a conversational agent online forms a relational scene together with surrounding people, work, emotions, and flows of information.

Forlano similarly draws on posthumanism to critique human-centered design, proposing a shift in attention away from the isolated individual user toward the relationships formed and negotiated within complex sociotechnical systems that encompass humans and nonhumans, individuals and organizations, and networks [14]. The designer's question shifts from what interface is for which user to what kinds of human-nonhuman relationships are assumed, and what interaction patterns and divisions of roles are made possible. Designing things with language capabilities thus involves imagining what positions and relationships those things will occupy as they generate conversations between people and their environments.

Some scholars connect this relational turn to questions of subjectivity. Braidotti, for instance, understands the subject not as an isolated individual but as a being constituted and transformed through multiple relationships [7]. Humans experience themselves through relationships with technologies, other people, and environments, sometimes taking on roles of care, at other times relying on others, and moving across positions. This perspective helps explain contemporary experiences in which emotions, desires, and patterns of action shift in response to relationships with technology. When we consider how some people perceive AI as a friend, others as a counselor, and others still as a teacher or authority, it becomes clear that the co-constitution of subjectivity and relationship is not merely an abstract claim.

What, then, do these complex theories and concepts collectively suggest, and why do they matter in a book composed of introductions to design artifacts? Put simply, the message we should take from them is one. To understand more clearly how AI affects human life and society, it is not enough to define what kind of technology AI essentially is. We need to carefully attend to how these technologies form relationships with humans in specific contexts, and what kinds of emotions, roles, and distances those relationships entail.

The studies presented in this book attempt to revisit encounters with voice-endowed things through this relational sensibility. As AI-enabled things that are perceived as humanlike become more common, the focus should not be on how well AI imitates humans. Living with these technologies is less about using new functions and more about learning unfamiliar relationships. The key questions concern the positions these technologies occupy in people's lives, the effects they have, and how they shape new forms of attachment, dependence, and mutual expectation across different social roles.

The challenge is that relationships are often felt bodily before they are articulated in words. Certain tones evoke closeness, specific responses convey authority, and inevitable silences heighten anxiety. To understand these shifts, it is therefore necessary not only to collect existing experiences but also to actively create and test possibilities for relationships that have not yet fully taken shape. In this spirit, rather than delving into conventional conversational agents themselves, I aimed to experimentally surface diverse relational possibilities between humans and AI-infused things. These experiments do not present a single correct answer to how AI should be designed. Instead, through concrete cases that suggest other imaginable relationships, they seek to foster new understanding and reflection on ways of living with AI. In the next chapter, I first outline the design research approach that enables this exploration, and then proceed to examine specific cases in turn.

References

1. Barad K (2007) Meeting the universe halfway: quantum physics and the entanglement of matter and meaning. Duke Up, Durham
2. Barrett LF (2017) How emotions are made: the secret life of the brain. Pan Macmillan, London
3. Bender EM, Koller A (2020) Climbing towards NLU: on meaning, form, and understanding in the age of data. In: Proceedings of the 58th annual meeting of the association for computational linguistics, pp 5185–5198
4. Berns K, Hirth J (2006) Control of facial expressions of the humanoid robot head ROMAN. In: 2006 IEEE/RSJ international conference on intelligent robots and systems. IEEE, New York, pp 3119–3124
5. Bohren J, Rusu RB, Jones EG, Marder-Eppstein E, Pantofar C, Wise M, Mösenlechner L, Meeussen W, Holzer S (2011) Towards autonomous robotic butlers: lessons learned with the PR2. In: 2011 IEEE international conference on robotics and automation. IEEE, New York, pp 5568–5575
6. Bown O (2012) Generative and adaptive creativity: a unified approach to creativity in nature, humans and machines. In: Computers and creativity. Springer, Berlin, pp 361–381
7. Braidotti R (2019) Posthuman knowledge, vol. 2. Polity Press, Cambridge
8. Burgard W, Cremers AB, Fox D, Hähnel D, Lakemeyer G, Schulz D, Steiner W, Thrun S (1999) Experiences with an interactive museum tour-guide robot. Artif Intell 114(1–2):3–55
9. Colomina B, Wigley M (2016) Are we human. In: Notes on an archaeology of design. Lars Müller, Zürich
10. Darling K (2016) Extending legal protection to social robots: the effects of anthropomorphism, empathy, and violent behavior towards robotic objects. In: Robot law. Edward Elgar Publishing, Cheltenham, pp 213–232
11. DiSalvo C, Gemperle F (2003) From seduction to fulfillment: the use of anthropomorphic form in design. In: Proceedings of the 2003 international conference on designing pleasurable products and interfaces, pp 67–72
12. Embgen S, Luber M, Becker-Asano C, Ragni M, Evers V, Arras KO (2012) Robot-specific social cues in emotional body language. In: 2012 IEEE RO-MAN: the 21st IEEE international symposium on robot and human interactive communication. IEEE, New York, pp 1019–1025
13. Farooq U, Grudin J (2016) Human-computer integration. Interactions, 23(6):26–32
14. Forlano L (2017) Posthumanism and design. She Ji: J Des Econ Innovation 3(1):16–29
15. Frauenberger C (2019) Entanglement HCI the next wave?. ACM Trans Comput-Hum Interact 27(1):1–27
16. Gerdes A (2016) The issue of moral consideration in robot ethics. ACM Sigcas Comput Soc 45(3):274–279
17. Haraway DJ (2003) The companion species manifesto: dogs, people, and significant otherness, vol. 1. Prickly Paradigm Press, Chicago, pp. 3–17
18. Haraway D (2013) A cyborg manifesto: science, technology, and socialist-feminism in the late twentieth century. In: The transgender studies reader. Routledge, Milton Park, pp 103–118

19. Harman G (2018) Object-oriented ontology: a new theory of everything. Penguin UK, London
20. Heidegger M (2010) Being and time. SUNY Press, New York
21. Hoffman G, Ju W (2014) Designing robots with movement in mind. J Hum-Rob Interact 3(1):91–122
22. Industrial Designers Society of America (2024) What is industrial design? https://www.idsa.org/about-idsa/advocacy/what-industrial-design/
23. Ihde D (1990) Technology and the lifeworld: from garden to earth (No. 560). Indiana University Press, Bloomington
24. Ji H, Han I, Ko Y (2023) A systematic review of conversational AI in language education: focusing on the collaboration with human teachers. J Res Technol Educ 55(1):48–63
25. Kuijer L, Giaccardi E (2018) Co-performance: conceptualizing the role of artificial agency in the design of everyday life. In: Proceedings of the 2018 CHI conference on human factors in computing systems, pp 1–13
26. Latour B (2005) Reassembling the social: an introduction to actor-network-theory. Oxford University Press, Oxford
27. McMillan D, Brown B, Kawaguchi I, Jaber R, Solsona Belenguer J, Kuzuoka H (2019) Designing with gaze: Tama–a gaze activated smart-speaker. Proc ACM Hum Comput Interact 3(CSCW):1–26
28. Meggs PB, Purvis AW, Maxa S, Sanders M (2025) Meggs' history of graphic design. Wiley, New York
29. Milne-Ives M, De Cock C, Lim E, Shehadeh MH, De Pennington N, Mole G, Normando E, Meinert E (2020) The effectiveness of artificial intelligence conversational agents in health care: systematic review. J Med Internet Res 22(10):e20346
30. Montemayor C, Halpern J, Fairweather A (2022) In principle obstacles for empathic AI: why we can't replace human empathy in healthcare. AI Soc 37(4):1353–1359
31. Moon A, Troniak DM, Gleeson B, Pan MK, Zheng M, Blumer BA, MacLean K, Croft EA (2014) Meet me where i'm gazing: how shared attention gaze affects human-robot handover timing. In: Proceedings of the 2014 ACM/IEEE international conference on Human-robot interaction, pp 334–341
32. Mumm J, Mutlu B (2011) Human-robot proxemics: physical and psychological distancing in human-robot interaction. In: Proceedings of the 6th international conference on Human-robot interaction, pp 331–338
33. Nass C, Steuer J, Tauber ER (1994) Computers are social actors. In: Proceedings of the SIGCHI conference on Human factors in computing systems, pp 72–78
34. Natarajan M, Gombolay M (2020) Effects of anthropomorphism and accountability on trust in human robot interaction. In: Proceedings of the 2020 ACM/IEEE international conference on human-robot interaction, pp 33–42
35. Nielsen J, Norman D (1998) The definition of user experience (UX). Nielsen Norman Group, London. https://www.nngroup.com/articles/definition-user-experience/
36. Niess J, Woźniak PW (2020) Embracing companion technologies. In: Proceedings of the 11th nordic conference on human-computer interaction: shaping experiences, shaping society, pp 1–11
37. Norman D (2013) The design of everyday things: revised and expanded edition. Basic books, New York
38. Perry A (2023) AI will never convey the essence of human empathy. Nat Hum Behav 7(11):1808–1809
39. Pradhan A, Findlater L, Lazar A (2019) "Phantom Friend" or "Just a Box with Information" personification and ontological categorization of smart speaker-based voice assistants by Older Adults. Proc ACM on Hum Comput Interact 3(CSCW):1–21
40. Rawal N, Stock-Homburg RM (2022) Facial emotion expressions in human–robot interaction: a survey. Int J Soc Robot14(7):1583–1604
41. Redström J, Wiltse H (2018) Changing things: the future of objects in a digital world. Bloomsbury Visual Arts

42. Rozendaal MC, Boon B, Kaptelinin V (2019) Objects with intent: designing everyday things as collaborative partners. ACM Trans Comput Hum Interact (TOCHI) 26(4):1–33
43. Rozendaal MC, Odom W, Marenko B (2021) Designing smart objects in everyday life: intelligences, agencies, ecologies. Bloomsbury Visual Arts
44. Scholl BJ, Tremoulet PD (2000) Perceptual causality and animacy. Trends Cogn Sci 4(8):299–309
45. Scott AE, Neumann D, Niess J, Woźniak PW (2023) Do you mind? User perceptions of machine consciousness. In: Proceedings of the 2023 CHI conference on human factors in computing systems, pp 1–19
46. Selwyn N (2019) Should robots replace teachers?: AI and the future of education. Wiley
47. Seymour W, Van Kleek M (2021) Exploring interactions between trust, anthropomorphism, and relationship development in voice assistants. Proc ACM on Hum Comput Interact 5(CSCW2):1–16
48. Shanahan M (2024) Talking about large language models. Commun ACM 67(2):68–79
49. Stuck RE, Rogers WA (2017) Understanding older adult's perceptions of factors that support trust in human and robot care providers. In: Proceedings of the 10th international conference on PErvasive technologies related to assistive environments, pp 372–377
50. Verbeek PP (2005) What things do: philosophical reflections on technology, agency, and design. Penn State Press, Pennsylvania
51. The Washington Post (2022) The Google engineer who thinks the company's AI has come to life. https://www.washingtonpost.com/technology/2022/06/11/google-ai-lamda-blake-lemoine/
52. Wakkary R (2021) Things we could design: for more than human-centered worlds. MIT Press, New York
53. Willoughby BJ, Dover CR, Hakala RM, Carroll JS (2025) Artificial connections: romantic relationship engagement with artificial intelligence in the United States. J Soc Pers Relat 42(12):3363–3387
54. Wittgenstein L (1998) Tractatus logico-philosophicus (C. K. Ogden, Trans.). Dover Publications
55. Zao-Sanders M (2025) How people are really using gen AI in 2025. In: Harvard business review. https://hbr.org/2025/04/how-people-are-really-using-gen-ai-in-2025

Chapter 2
Design as a Means of Inquiry

Abstract In this chapter, I outline the design research approach I employ in this book within the contexts of research through design and speculative design. To help readers understand the three case studies presented later, not as mere design projects or artistic outputs, but as research artifacts for exploring human-AI relationships, I outline how design has produced knowledge within human-computer interaction (HCI). First, I discuss how design artifacts have expanded beyond being mere containers of knowledge. They have become means for constructing new questions and opening interpretive spaces through the design process and iterative production and use. Next, it examines how design, through alternative approaches like critical, frictional, and speculative design, has positioned itself as a discursive practice that reflects on sociotechnical realities and imagines possible futures, moving beyond being merely a tool for problem-solving or improvement. Finally, to complement the limitations of design fiction, it outlines how approaches involving actual use and materiality draw imagination into experiential events and the friction of everyday life. This clearly positions the case studies presented in this book not as definitive answers, but as exploratory research that opens possibilities.

2.1 The Relationship Between Design and Knowledge

Design has long occupied a distinctive and important position in human-computer interaction (HCI) because of its fundamentally generative and creative character [22]. In domains that engage with emerging technologies in particular, design's capacity to reflect on subjective and complex human experiences and to translate them into material or conceptual outcomes often plays a central role [19, 41]. Design researchers have emphasized the importance of actively involving stakeholders in the design process [15, 29] and have introduced techniques such as personas to deepen contextualized understanding of specific user groups [5, 6]. Through these efforts, discussions of how technologies should be designed have increasingly taken shape in ways that consider technology and humans in a more balanced manner.

As the number of designers and researchers with design backgrounds within the HCI community has grown, so too have discussions on how the outcomes of design

H. Cho, *The AI as Social Actor*, KAIST Research Series,
https://doi.org/10.1007/978-981-92-0537-0_2

activity can produce forms of knowledge that meet the expectations of an academic community [25, 34, 47].

Notably, Pierce organizes the relationship between design artifacts and knowledge through three perspectives, in which artifacts consist of, contain, or construct knowledge [34]. Early design research often treated artifacts as final implementations that carried knowledge [21]. In contemporary HCI, however, it is widely accepted, following Cross, that design processes themselves are means of knowledge production and that designed artifacts contain knowledge [7]. From this perspective, interaction designers generate a distinctive form of knowledge by exploring not only how to solve a given problem but also what constitutes a problem and what should be designed. Scholars have also emphasized that the nonlinear and iterative nature of design research produces new types of knowledge [9, 32].

Another perspective treats artifacts as tools that construct knowledge [20, 27]. Here, the artifact is understood not as the endpoint of research, but as a core means of knowledge building and as an intermediate position closer to a point of departure. Artifacts can be used to explore the potential value of technologies [36, 37], to mediate philosophical inquiry [18, 26], or to reframe problems from entirely different perspectives [17].

While scholars differ in what they emphasize, research through design (RtD), one of the most widely referenced approaches in contemporary HCI [22, 43, 47], accommodates these perspectives in a relatively broad manner. RtD understands design practice not simply as an activity that proposes solutions to predefined problems, but as a methodological framework for producing knowledge through design processes and their outcomes. The cases discussed in this book are situated within this RtD tradition. Through the formation and use of artifacts, and through the stories that surround them, they seek to develop new ways of understanding human-AI relationships.

2.2 The Emergence and Expansion of Speculative Design

Over time, the role of design artifacts in research has continued to expand. Today, they are frequently used as tools to stimulate critical thinking and encourage reflective discourse on technological, ethical, and social issues. As discussed by Pierce in relation to the concept of *frictional design*, this approach goes beyond the productive values often expected by research communities and audiences, such as improving something or producing better outcomes [35]. Frictional design challenges productivity-oriented notions of progressive design by deliberately creating discomfort and tension, posing different questions about technology and society.

Critical design is a representative example. It aims to prompt people to reflect on assumptions taken for granted in everyday life through questions intentionally posed by designers [12, 13]. While it originated as a critique of industrial design's role in reinforcing capitalist consumer culture, within HCI it has evolved into a way of exploring ethical and social issues surrounding technology and of opening new research areas [1, 39].

Along similar lines, other researchers have proposed various goals and attitudes for design research. *Adversarial design* [11] and *reflective design* [40] draw attention to the stance designers take and to the politics of intervention. In contrast, approaches such as *slow design* [30, 33] and *ludic design* [23] reconsider values that had been overlooked in earlier discourses centered on efficiency and productivity, including slowness, leisure, and play. There are also many conceptual variations of alternative design approaches, including *speculative design* [14], *reflective design* [40], and *discursive design* [45]. Although each has its own characteristics, they are loosely connected by a shared commitment to reflecting on current practices and exploring new possibilities. In the literature, they are often used in relation to one another rather than being sharply distinguished [1, 35].

In recent years, the term "speculative design" has tended to be preferred as a relatively broad umbrella concept. I also often position my own work as part of speculative design, for two main reasons. First, speculative design is generally less confrontational. Rather than engaging primarily in deep theoretical debate aimed at a narrow group of experts, it tends to invite broader participation by allowing people to enter discussions through pleasurable acts of imagination. This aligns with my own research preferences. Second, speculative design is future oriented. It focuses on exploring potential futures grounded in technological possibilities. That is, it centers on posing "What if?" questions mediated by emerging technology, which closely resonates with my research interests.

In speculative design practice, design is understood as an activity that generates alternative possibilities to reflect on the present critically. By designing fictional situations and things that have not yet been realized or do not exist, it encourages creative and critical discourse around norms and realities that are otherwise taken for granted. For example, Søndergaard and Hansen proposed devices that combine intimate voices with feminist perspectives to surface gendered ethical and philosophical issues that might arise when digital personal assistants enter the home [42]. Another study combined IoT-based domestic data with first-person narratives to explore what kinds of intimacy and burden are produced by living with data that is always present [8]. Other work has examined the social implications of using drones in policing through a game format [28], or imagined future university campuses grounded in sustainability [25]. A wide range of topics has been addressed through speculative approaches.

As these examples show, alternative design approaches do not limit their purpose to proposing concrete methods for making society more preferable, nor to offering actionable design recommendations for designing the right things. Instead, they reposition design as an activity that critically reflects on technology and generates discourse around the directions of technological development and the forms of life that follow from it [1, 4].

2.3 Speculating with Tangible Artifacts

One limitation of speculative research is that, because it deals with future or fictional situations, it often relies heavily on mediated representations such as text, images, or video. While such design fictions are compelling for sharing ideas and activating conversation [2, 3, 44], they struggle to fully reveal elements that emerge from interaction grounded in physical experience, such as sensation, rhythm, hesitation, and habit. In response to these limitations, HCI design research has developed ways of drawing imagination into experienceable events or into the materiality of everyday life.

Speculative enactment, for instance, explores imagined futures not by merely thinking about or reading them, but by having participants enact those possibilities within their present lives [16]. This approach is compelling because it pulls abstract possibilities into present sensation, revealing what feels awkward, what unfolds smoothly, and what new meanings emerge when people attempt to live with them. Within this methodology, researchers take seriously the task of designing conditions that encourage participants to invest genuinely in alternative scenarios.

An approach that resonates even more directly with the studies presented in this book is material speculation [46], which treats actual and situated artifacts as sites of critical inquiry. Wakkary positions counterfactual artifacts as a complementary concept to design fiction, arguing that everyday material experiences themselves can be designed to generate critical questions. The intention is to treat moments in which the things we encounter, touch, and live with in daily life rub against existing assumptions as opportunities to spark new imaginaries.

This approach gains credibility because there is already a substantial body of research in which artifacts form part of the argument itself. Photobox, for example, was designed to print Flickr images intermittently over long periods, transforming digital photos from instantly consumed information into a medium through which relationships are re-formed as time accumulates [31]. Prayer companion materializes streams from Really Simple Syndication (RSS) feeds or social networking services, reorganizing the practice of deciding what to pray for and prompting spiritual and social reflection on how closed communities might connect with the outside world [24]. Odd Interpreters translates smart home data into imaginative and sensorial forms, rather than precise dashboards, reopening the experience of living with data by pluralizing encounters rather than linear control [10]. Obscura 1C Camera, which requires users to physically destroy the camera in order to access the photos they have taken, encourages reflection on the value and controllability of owning digital data [37, 38]. Compared to ideas that exist only in text, such devices enable much more immediate and tangible forms of reflection.

The series of studies introduced in this book is likewise grounded in the RtD tradition and in the philosophy of speculative design. I adopt an approach that utilizes artifacts to reveal the roles AI-infused things play in everyday life and the effects they have on us. More specifically, I attend to how people's expectations, trust, dependence, distance, and attachment are configured in diverse ways through these processes. For this reason, it is most appropriate to view the cases in this book not as providing a single answer to what ideal human-technology relationships should be, but as attempts to open up new reflections and discussions on how we live with AI by examining these relationships from fresh perspectives.

References

1. Bardzell J, Bardzell S (2013) What is "critical" about critical design?. In: Proceedings of the SIGCHI conference on human factors in computing systems, pp 3297–3306
2. Bleecker J (2009) Design Fiction: a short essay on design, science, fact, and fiction. Near Future Laboratory. https://nearfuturelaboratory.com/blog/2009/03/design-fiction-a-short-essay-on-design-science-fact-and-fiction/. Accessed 16 May 2026
3. Blythe M (2014) Research through design fiction: narrative in real and imaginary abstracts. In: Proceedings of the SIGCHI conference on human factors in computing systems, pp 703–712
4. Candy S, Dunagan J (2017) Designing an experiential scenario: the people who vanished. Futures, 86:136–153
5. Chang YN, Lim YK, Stolterman E (2008) Personas: from theory to practices. In: Proceedings of the 5th Nordic conference on Human-computer interaction: building bridges, pp 439–442
6. Cooper A (1999) The inmates are running the asylum. In: Software-ergonomie'99: design von informationswelten. Vieweg+ Teubner Verlag, Wiesbaden, pp 17–17
7. Cross N (1999) Design research: a disciplined conversation. Design Issues 15(2):5–10
8. Desjardins A, Biggs HR (2021) Data epics: embarking on literary journeys of home internet of things data. In: Proceedings of the 2021 CHI conference on human factors in computing systems, pp 1–17
9. Desjardins A, Key C (2020) Parallels, tangents, and loops: reflections on the 'Through' part of RtD. In: Proceedings of the 2020 ACM designing interactive systems conference, pp 2133–2147
10. Desjardins A, McWhirter J, Petelka J, Simon C, Shin Y, Peven RK, Widjaja P (2023) On the making of alternative data encounters: The odd interpreters. In Proceedings of the 2023 CHI conference on human factors in computing systems, pp 1–20
11. DiSalvo C (2015) Adversarial design. MIT Press, Cambridge
12. Dunne A (2008) Hertzian tales: electronic products, aesthetic experience, and critical design. MIT Press, Cambridge
13. Dunne A, Raby F (2021) Design noir: the secret life of electronic objects, vol 2. Bloomsbury Publishing, London
14. Dunne A, Raby F (2024) Speculative everything, with a new preface by the authors: design, fiction, and social dreaming. MIT Press, Cambridge
15. Ehn P (2008) Participation in design things. In: Participatory design conference (PDC), Bloomington, Indiana (2008). ACM Digital Library, pp 92–101
16. Elsden C, Chatting D, Durrant AC, Garbett A, Nissen B, Vines J, Kirk DS (2017) On speculative enactments. In: Proceedings of the 2017 CHI conference on human factors in computing systems, pp 5386–5399
17. Encinas E, Blythe M, Lawson S, Vines J, Wallace J, Briggs P (2018) Making problems in design research: the case of teen shoplifters on Tumblr. In: Proceedings of the 2018 CHI conference on human factors in computing systems, pp 1–12
18. Encinas E, Durrant AC, Mitchell R, Blythe M (2020) Metaprobes, metaphysical workshops and sketchy philosophy. In: Proceedings of the 2020 CHI conference on human factors in computing systems, pp 1–13

19. Forlizzi J, DiSalvo C, Bardzell J, Koskinen I, Wensveen S (2011) Quality control: a panel on the critique and criticism of design research. In: CHI'11 extended abstracts on human factors in computing systems, pp 823–826
20. Forlizzi J, Zimmerman J, Hekkert P, Koskinen I (2018) Let's get divorced: constructing knowledge outcomes for critical design and constructive design research. In: Proceedings of the 2018 ACM conference companion publication on designing interactive systems, pp 395–397
21. Frayling C (1994) Research in art and design. Royal College of Art Research Papers, vol 1, no 1, 1993/4
22. Gaver W (2012) What should we expect from research through design?. In: Proceedings of the SIGCHI conference on human factors in computing systems, pp 937–946
23. Gaver WW, Bowers J, Boucher A, Gellerson H, Pennington S, Schmidt A, Steed A, Villars N, Walker B (2004) The drift table: designing for ludic engagement. In: CHI'04 extended abstracts on Human factors in computing systems (pp. 885–900).
24. Gaver W, Blythe M, Boucher A, Jarvis N, Bowers J, Wright P (2010) The prayer companion: openness and specificity, materiality and spirituality. In Proceedings of the SIGCHI conference on Human factors in computing systems, pp 2055–2064
25. Hauser S, Desjardins A, Wakkary R (2014) Sfuture: envisioning a sustainable university campus in 2065. In: Proceedings of the 2014 companion publication on Designing interactive systems, pp 29–32
26. Hauser S, Oogjes D, Wakkary R, Verbeek PP (2018) An annotated portfolio on doing post-phenomenology through research products. In: Proceedings of the 2018 designing interactive systems conference, pp 459–471
27. Koskinen I, Zimmerman J, Binder T, Redstrom J, Wensveen S (2013) Design research through practice: from the lab, field, and showroom. IEEE Trans Profess Commun 56(3):262–263
28. Lindley J, Coulton P (2015) Game of drones. In: Proceedings of the 2015 annual symposium on computer-human interaction in play, pp 613–618
29. Muller MJ, Kuhn S (1993) Participatory design. Commun ACM 36(6):24–28
30. Odom W, Banks R, Durrant A, Kirk D, Pierce J (2012) Slow technology: critical reflection and future directions. In: Proceedings of the designing interactive systems conference, pp 816–817
31. Odom WT, Sellen AJ, Banks R, Kirk DS, Regan T, Selby M, Forlizzi JL, Zimmerman J (2014) Designing for slowness, anticipation and re-visitation: a long term field study of the photobox. In: Proceedings of the SIGCHI conference on human factors in computing systems, pp 1961–1970
32. Odom W, Wakkary R, Lim YK, Desjardins A, Hengeveld B, Banks R (2016) From research prototype to research product. In: Proceedings of the 2016 CHI conference on human factors in computing systems, pp 2549–2561
33. Odom W, Stolterman E, Chen AYS (2022). Extending a theory of slow technology for design through artifact analysis. Human–Comput Interact 37(2):150–179
34. Pierce J (2014) On the presentation and production of design research artifacts in HCI. In: Proceedings of the 2014 conference on designing interactive systems, pp 735–744
35. Pierce J (2021) In tension with progression: grasping the frictional tendencies of speculative, critical, and other alternative designs. In: Proceedings of the 2021 CHI conference on human factors in computing systems, pp 1–19
36. Pierce J, Paulos E (2014) Counterfunctional things: exploring possibilities in designing digital limitations. In: Proceedings of the 2014 conference on Designing interactive systems, pp 375–384
37. Pierce J, Paulos E (2015) Making multiple uses of the obscura 1C digital camera: reflecting on the design, production, packaging and distribution of a counterfunctional device. In: Proceedings of the 33rd annual ACM conference on human factors in computing systems, pp 2103–2112
38. Pierce J, Paulos E (2015) Obscura 1C digital camera. In: Proceedings of the 2nd biennial research through design conference
39. Rosner DK (2018) Critical fabulations: reworking the methods and margins of design. MIT Press, Cambridge

40. Sengers P, Boehner K, David S, Kaye JJ (2005) Reflective design. In: Proceedings of the 4th decennial conference on Critical computing: between sense and sensibility, pp 49–58
41. Smith GC, Tabor P (1996) The role of the artist-designer. In: Bringing design to software, pp 37–61
42. Søndergaard MLJ, Hansen LK (2018) Intimate futures: staying with the trouble of digital personal assistants through design fiction. In: Proceedings of the 2018 designing interactive systems conference, pp 869–880
43. Stappers P, Giaccardi E (2014) 43. Research through Design. The encyclopedia of human-computer interaction. Carnegie Mellon University, Pittsburgh
44. Sterling B (2009) Cover story design fiction. Interactions 16(3):20–24
45. Tharp BM, Tharp SM (2019) Discursive design: critical, speculative, and alternative things. MIT Press, Cambridge
46. Wakkary RL, Odom WT, Hauser S, Hertz G, Lin H (2016) Material speculation: actual artifacts for critical inquiry. In: 5th decennial aarhus conference on critical alternatives August (AA 2015). Aarhus University, pp 97–108
47. Zimmerman J, Forlizzi J, Evenson S (2007) Research through design as a method for interaction design research in HCI. In: Proceedings of the SIGCHI conference on Human factors in computing systems, pp 493–502

Part II
Design Speculations

How might AI and humans form relationships in diverse ways? This part introduces three speculative design cases, Beau, Areca, and ShamAIn, in order to provoke discussion around this question. Before turning to these cases, I first present a relational map that I have developed. This map serves to secure a broader range of relational possibilities within the constraints of a limited research series.

Chapter 3
A Temporary Framework for Exploring Human-AI Relationships

Abstract This chapter presents a provisional framework for reading the three design cases introduced in Part II. As AI permeates things, the human-thing relationship can no longer be neatly summarized by the single term tool. Even the same technology can feel intimate depending on its context and the role users expect it to play, maintaining distance as a social entity or sometimes occupying a position of authority. Rather than creating a classification system that inherently explains this diversity, this chapter proposes a framework as a thinking tool to illustrate how different relational configurations contrast, even within a limited exploration. The frame consists of two axes. First, relational distance indicates the degree of intimacy and emotional closeness. Second, relational hierarchy focuses on how responsibility and initiative are distributed within the relationship. These axes draw inspiration from interpersonal relationship theory. However, they are considered simplified devices, not to equate human-AI relationships with interpersonal ones, but to capture in explainable language the sense of relationship that transcends the tools we actually experience today. Finally, this chapter positions three design research cases, Beau, Areca, and ShamAIn, at different points within this framework, foreshadowing the directions of relational possibilities that subsequent chapters will explore.

3.1 Mapping the Relationships

In Part I, I examined how AI enables us to experience things as more than mere tools, and how this is shifting our relationship with technology toward a more relational direction. AI-embedded things, in particular, make it difficult to describe the relationship between humans and things within a single category.

In some cases, AI is experienced as a close presence, while in others, it functions as an entity with a public role or authority. In yet other scenarios, it is still clearly perceived as a tool. Crucially, this difference cannot be attributed solely to the type of technology. Even the same technology can have a vastly different relational character depending on the context in which it is placed and the expectations people bring to their interaction with it.

H. Cho, *The AI as Social Actor*, KAIST Research Series,
https://doi.org/10.1007/978-981-92-0537-0_3

Before exploring these complex and diverse relationships, I begin by establishing a space where different relational possibilities can be laid out side by side to reveal how they emerge and contrast with one another. This space is not a taxonomy explaining the essence of human-AI relationships. It is, quite literally, a temporary framework as a tool for thought, designed to make the diversity of relational configurations comparable within limited resources for exploration. Rather than attempting to implement every possible relationship exhaustively, I deliberately arrange relationships in distinct directions to more clearly reveal diverse questions.

This coordinate system has two axes. The first is relational distance, and the second is relational hierarchy. These two axes draw inspiration from various theories developed to understand human relationships, but they are simplified for this book. To reiterate, these axes are not an attempt to equate human-AI relationships with interpersonal ones. Instead, they are metaphorical devices provisionally borrowed to capture, in explainable language, the sense of the beyond-tool-relationship we actually experience now that things increasingly speak, respond, and perform roles.

3.2 Relational Distance

Relational distance refers to the degree of psychological intimacy and emotional closeness one feels toward another being. This concept draws inspiration from Edward Hall's proxemics discussions, which categorized the use of space between people according to levels of intimacy and trust. Hall divided interpersonal distance into intimate, personal, social, and public zones, demonstrating that distance is not merely physical separation but intertwined with modes of emotional and social connection [5]. This book simplifies this into three levels, tailored to exploring human-thing relationships.

- **Intimate:** Intimate relationships involve profound emotional bonds, typically seen with close family members, romantic partners, or trusted friends. Such relationships are characterized by trust, understanding, and significant life influence. Technological things deeply integrated into personal lives could foster similar emotional bonds in the future. For instance, systems capable of offering emotionally sensitive responses or managing users' life patterns with precision may encourage intimate connections.
- **Sociable:** Sociable relationships prioritize collaboration and interaction over emotional closeness, resembling relationships with friends or colleagues. These interactions are built on trust and cooperation, rather than deep emotional connections. Things designed for collaborative or playful functionalities could support sociable relationships. For example, technologies that enhance teamwork or enable interactive experiences might cultivate such bonds.
- **Formal:** Formal relationships are goal oriented and pragmatic, often defined by specific purposes, such as professional exchanges or brief transactions. These relationships are more functional and transient, with limited emotional

involvement. Things used in task-specific or short-term contexts may align with this category. For example, devices designed for straightforward functions in public or professional settings could exemplify formal relationships.

Relational distance helps reveal that human-thing relationships possess a texture not captured by the frequency or duration of interaction. This is because some things can seem distant, even when they are frequently present, while others can feel intensely close at specific moments, despite being used infrequently.

3.3 Relational Hierarchy

Relational hierarchy addresses how roles and initiative are distributed within relationships. This draws on relational theories that have long noted relationships between people are not always equal, but are sometimes organized around authority and direction, sometimes around cooperation and equality, and sometimes around submission and delegation. For instance, Fiske's Relational Models Theory explains that human relationships are structured in various ways, including models based on authority and hierarchy [6]. This book simplifies them into three levels for human-thing relationships.

- **Superior:** Superior relationships arise when one entity possesses greater authority or capability than another. These relationships, akin to those between a coach and players or a teacher and students, emphasize the centrality of the authoritative party's decisions. Future things might assume superior roles, offering guidance or making decisions in specific contexts.
- **Equivalent:** Equivalent relationships are characterized by equality in status and shared roles, similar to relationships between peers. These interactions emphasize mutual trust and cooperation. Design approaches promoting equivalence could position objects as collaborative partners. For instance, systems designed to work alongside humans toward shared objectives might establish equivalent relationships.
- **Subordinate:** Subordinate relationships occur when one entity is subject to another's authority, such as in supervisor-subordinate dynamics. These relationships are defined by the subordinate's compliance with the superior's directives. Things occupying subordinate roles may function as tools that strictly follow human instructions, such as executing repetitive tasks or responding to commands.

Hierarchy is constructed not by AI's inherent capabilities but by how people perceive responsibility and judgment within interactions, and who they feel determines the direction of the relationship. Therefore, hierarchy should be understood as another sensory indicator of how relationships form, distinct from measures of technological performance.

3.4 Propositions of Things

The three design cases in this book are structured to experiment with different relational sensibilities within the above framework (see Fig. 3.1). Each begins at a different point of relational distance while simultaneously revealing a distinct relational hierarchy. Through these cases, I sought not to ask "How humanlike can AI make things?" but rather "How can AI reorganize the relationship between humans and things?" Here, I first briefly introduce the three things positioned at different points on this framework. Subsequent chapters will present these as research, focusing on aspects such as design decisions and their underlying rationale, the progression of the study, and an analysis of the empirical data. To pique readers' interest, I introduce each here in a more artistic tone.

3.4.1 Beau

In crowded spaces, robots always become ambiguous beings. They stand with conspicuous bodies, yet a layer of hesitation remains before they are naturally accepted as being there like anyone else. Passersby glance at them. They briefly ponder whether to take a photo, strike up a conversation, or simply pretend not to notice. If the robot initiates by waving or mimicking a smile, that hesitation

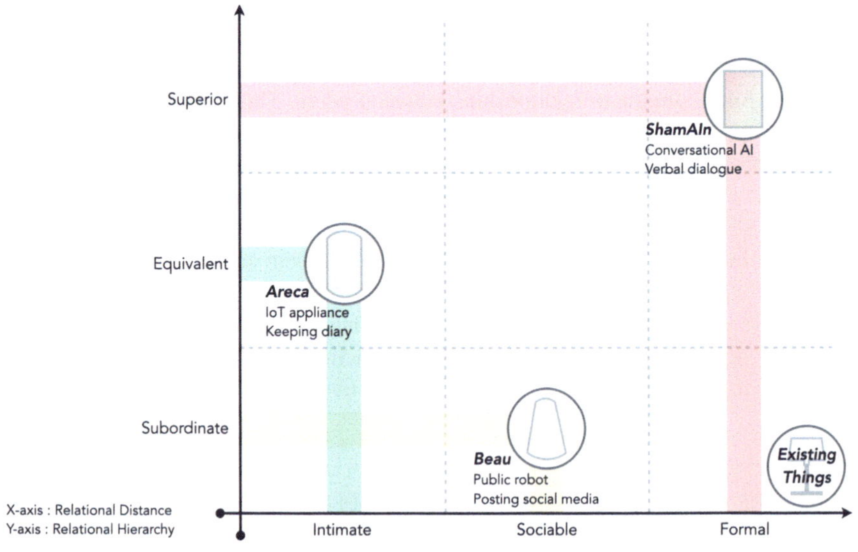

Fig. 3.1 A temporary relational framework mapping the three design cases explored in this book. The horizontal axis represents relational distance, while the vertical axis represents relational hierarchy. Beau, Areca, and ShamAIn occupy distinct regions of this space

momentarily eases. Soon after, another awkwardness follows. Was that smile truly meant for me? Should I respond politely now? Or am I just interpreting its programmed actions arbitrarily?

Beau is a robot that embraces this public awkwardness head-on [4]. However, Beau's way of connecting with people is not through the kind of intricate expressions or gestures one might typically imagine. Rather than putting its body forward, Beau initiates conversation through text. This text is not a whisper audible only to a select few. Instead, Beau reveals itself through a format people already associate with reading others, namely, social media posts. We hesitate to start conversations with strangers suddenly, yet we relatively easily engage with reading someone's post on our timeline. Beau subtly rides that inertia. Moreover, as is often the case with relationships in public spaces, it prioritizes sociability over intimacy. It offers a proposal such as "We briefly crossed paths, and you passed me by, but I can turn this moment into a story."

That does not mean Beau takes the lead in the relationship. Beau asks people to read, to react, and to share a moment of attention. The key to opening the door of the relationship remains in human hands. Regardless of social media posts, on-site Beau is just a robot helping with temperature checks. In that sense, Beau is clearly subordinate in its relationship with humans.

Nevertheless, precisely because of this subordinate position, the robot comes across not as a threat but as a light social pressure. It becomes a presence that does not get too close, yet feels slightly uncomfortable to ignore completely. That unique, shallow sense of responsibility we feel toward each other in public spaces, along with a layer of politeness that is neither overly warm nor cold, shapes Beau's relational sensibility.

The question this case raises is not simply "Can robots be social?" It is more specific. How can a thing in a public space request human attention? And when that request succeeds, do we still perceive it solely as a machine's output, or do we experience it as someone's presence? Rather than asking whether things in public spaces must mimic humans to form social relationships, Beau questions whether there might be another grammar for designing sociable distance.

3.4.2 Areca

Things in the home often have a surprisingly strong presence even without speaking. This is especially true for devices that are always in place and occasionally signal their status through lights or sounds. Air purifiers, in particular, are often left running and forgotten. One day, however, we suddenly realize we have entrusted them with too much.

Areca explores precisely that point, the conditions of intimacy that arise when things deeply enter our daily lives. However, Areca does not create this sense of intimacy by exaggerating emotional expressions or behaving affectionately like a human being. Instead, it remains quiet and restrained. It encourages people to feel that time passes for it too, that it has its own record of moments, and that experiences accumulate through observation and recording. The writing format resembles a personal record rather than one intended for public display [1, 2]. This subtle shift gradually alters how people interact with the device. Their perspective shifts from "operating this device" toward "living alongside this device."

What matters in an intimate relationship is often the sensations that are shared. When changes in the air, the rhythm of the home, or the user's daily patterns become woven into Areca's day, the user finds it difficult to feel complete control over the device's life. Technically, we turn it on and off. Experientially, something more unfolds. The thing assists us, and at the same time, we adapt to its rhythm. The relationship cannot be defined solely through hierarchy. Areca remains a subordinate tool that executes commands, yet it simultaneously feels like an equal partner that co-creates part of the living environment. This sense of equality does not stem from humanlike qualities. Instead, it emerges because roles within the relationship do not move in only one direction. Moments arise when we care for Areca and when we feel Areca cares for us. As this exchange repeats, intimacy gradually takes hold.

This case raises a question. What power do the words of things possess at an intimate distance? Words are not merely a means of conveying information. They shape the rhythm of a relationship, reconfigure senses of responsibility, and influence how thoughts and actions unfold. Areca was designed with the idea of a machine having an inner life, yet the authenticity of that inner life is not the focus here. Instead, this case observes how intimacy and equality emerge when quiet domestic things express themselves through diary-like texts, and how such relationships leave a particular texture on human habits and emotions.

3.4.3 ShamAIn

We often expect our relationship with AI to revolve around comfort. Yet some social relationships demand dignity before comfort. In these relationships, the weight of a single word differs, the pressure of silence feels heavier, and the attitude one must adopt is clearly defined. Religious or ceremonial moments provide familiar examples. In such scenes, what matters is not the accuracy of information or emotional empathy. Instead, it is the authority and interpretive gaps produced through language. People locate meaning within those gaps, justify decisions, and sometimes persuade themselves to act.

ShamAIn explores precisely this type of relationship through an interactive AI format. Here, the AI does not appear as a friend-like conversational partner. Instead, it responds from a position resembling a higher being within a shamanistic space. Its tone avoids familiarity and relies on ritualistic rhythms and symbolic structures that resist easy consumption [3]. This configuration reshapes the structure of interaction. People ask questions with seriousness, linger over the responses, interpret them, and attempt to connect them to their own lives. The exchange becomes less about information transfer and more about interpretation.

Within this setting, the hierarchy of the relationship becomes apparent. ShamAIn appears to stand above the user's judgment by offering direction rather than serving as a subordinate tool. Even when people know the responses are algorithmically generated, they often struggle to dismiss them outright. This elevated position does not imply that the AI holds actual power. Instead, it reflects the user's tendency to receive its words as if they carried authority. At times, people want decisions to be made on their behalf. At other times, they want words that carry enough weight to confirm their own choices. At times, recognition from an authority figure feels more compelling than comfort from a peer. ShamAIn suggests that AI can occupy this role.

The question this case raises is stark. When does a language-enabled thing move beyond being a counseling tool and come to be experienced as an authority figure? When such an experience becomes possible, what responsibilities do people assign to its words, and which responsibilities do they quietly avoid? How easily does such a relationship become captivating? ShamAIn is not intended to evaluate AI's persuasiveness or accuracy. Instead, it demonstrates how conversational form, symbolic arrangement, and the rhythm of speech can work together to construct formal distance and superior hierarchy. Through this construction, the case examines how AI can impact human life and the tensions and temptations that such interventions produce.

References

1. Cho H, Lee J, Ku B, Jeong Y, Yadgarova S, Nam TJ (2023) ARECA: a design speculation on everyday products having minds. In: Proceedings of the 2023 ACM designing interactive systems conference, pp 31–44
2. Cho H, Nam TJ (2025) Living alongside areca: exploring human experiences with things expressing thoughts and emotions. In: Proceedings of the 2025 CHI conference on human factors in computing systems, pp 1–16
3. Cho H, Seo JA, Lee J, Kim CM, Nam TJ (2025) ShamAIn: designing superior conversational AI inspired by shamanism. In: Proceedings of the 2025 CHI conference on human factors in computing systems, pp 1–18
4. Cho H, Nam TJ (2023) The story of Beau: exploring the potential of generative diaries in shaping social perceptions of robots. Int J Design 17(1):1–15
5. Hall ET, Birdwhistell RL, Bock B, Bohannan P, Diebold Jr. AR, Durbin M, Edmonson MS, Fischer JL, Hymes D, Kimball ST, La Barre W (1968) Proxemics [and comments and replies]. Curr Anthropol 9(2/3):83–108
6. Haslam N, Fiske AP (1999) Relational models theory: a confirmatory factor analysis. Pers Relationships 6(2):241–250

Chapter 4
The Sociable Subordinate: Public Robot Beau

Abstract This chapter introduces Beau, a public robot that uploads daily text posts through a pseudo-social media platform. I describe the design and implementation of a series of research prototypes that explore the experience of interacting with a robot through social media posts and how such interactions shape perceptions and experiences of the robot. The prototypes consist of three components: Beau, HeyBeau, and IamBeau. Beau is a public robot that measures visitors' body temperature at a building entrance and shares text posts written from the robot's perspective with users. Because the study took place during the COVID-19 pandemic, Beau was used very frequently. HeyBeau is a curation system designed to support valid and effective user studies. It allows researchers to select appropriate topics from the data collected by Beau and to review automatically generated posts for contextual errors or ethical concerns before they are shown to users. Finally, IamBeau is a social media-style Web application that enables people to read these automatically generated posts. These three elements were carefully designed with attention to appearance, a realistic research context, and data collection practices that respect privacy. During a 16-day field deployment, 12 participants read a total of 32 robot perspective posts through IamBeau. Changes were tracked through four administrations of the RoSAS questionnaire and post-study semi-structured interviews. Quantitative analysis showed that robot perspective text significantly increased perceived Warmth among the robot's social attributes, while no significant changes were found for Competence or Discomfort. Qualitative analysis revealed that these changes were linked to the way intelligence, consciousness, and emotion were expressed through text, how accumulated posts came to constitute the robot's identity, and how users began to accept the robot as an acquaintance or a potential friend. At the same time, the study identified mismatches that arise when expectations generated by text are not fully met through physical interaction, tensions between the efficiency of automatic generation and the demands of authenticity and ethical filtering, and the need for deliberate data and model design to shape social perception intentionally. This chapter suggests that when a thing generates and shares its own narrative through an LLM, relationships between humans and things can become more sociable.

H. Cho, *The AI as Social Actor*, KAIST Research Series,
https://doi.org/10.1007/978-981-92-0537-0_4

4.1 Backgrounds and Research Goal

Robots are perhaps the most representative case of technological anthropomorphism. They move on their own, and in many cases, their very purpose is to perform roles associated with humans. For this reason, researchers have long agreed that social interaction and perceptions of robot sociability are essential considerations for designers [11]. Traditionally, much research has focused on enhancing a robot's social presence through physical design features or nonverbal cues such as gestures, facial expressions, and gaze [1, 15, 28]. These approaches aim to meet user expectations by anthropomorphizing robots to behave like humans or by designing them to foster emotional bonds similar to those formed with pets [8, 10]. However, robot designs that rely primarily on appearance and limited physical cues can lead to predictable and repetitive interactions, which may ultimately reduce user interest [12, 16]. Research on interaction elements that make robots more engaging has therefore been extensive. In short, robots have long been regarded as things that can form less formal and more sociable relationships than ordinary things.

These studies have steadily expanded our understanding of what it means to design social robots. On a more personal level, however, I have come to believe that whether an entity is sociable depends crucially on whether it possesses a complex narrative. This includes the kinds of experiences it has had, its personality, and the values it holds. Such qualities make each entity unpredictable and singular. Paradoxically, this unpredictability may be precisely what allows it to be social. This is not a question that can be resolved within a single study, and it remains an ongoing concern in my work. Two points are especially important. First, research on social robots has primarily focused on interactions with humans. Second, designers have tended to define robot interactions in advance, transparently, in terms of appearance or behavior. This is a reasonable approach from the standpoint of usability. Yet what contributes to deeper social relationships may be the inherently opaque emotions or thoughts that belong to the entity itself [7, 10].

In this study, I explore the possibility that a robot can express its own experiences, thoughts, and emotions using an LLM and, in doing so, construct its own narrative. The aim is not simply to avoid clashes with users' initial expectations, but to actively shape users' social perceptions of the robot. Through a case study of the public service robot Beau, I conduct an initial exploration of whether generated diaries can enhance perceived robot sociability and how they might lead to different forms of relationships between humans and robots.

In summary, this research is guided by two questions:

- Can a thing be designed to be sociable in its relationship with humans through its own narrative?
- How do humans interact with such a thing, and how do they form relationships with it?

4.2 Design Rationales, Decisions, and Outcomes

To address the research questions outlined earlier, I designed Beau, a robot that posts about its own experiences on social media. This design includes not only the robot itself but also a human-in-the-loop system for generating social media posts and an interface through which people can view the uploaded content. This section describes the design rationales, implementation process, and final form of the research prototype system composed of Beau, HeyBeau, and IamBeau.

4.2.1 Beau

Beau is a public service robot that greets visitors at the entrance of a building (Fig. 4.1). Its primary function is to measure visitors' body temperature. This choice offers two important advantages. First, during the COVID-19 pandemic, temperature checks were mandatory for entry into public spaces. Since this study was conducted during that period, it allowed for the observation of interactions between the robot and a wide range of people. Second, the repetitive nature of temperature measurement creates conditions that allow people's experiences to be sufficiently reflected upon over time. For these reasons, I considered a temperature-measuring robot to be the most suitable prototype for this study. This stands in contrast to guide robots that are only needed by some people, or cleaning robots that involve minimal interaction with users.

In designing Beau, I considered prior work on how different interaction modalities influence social perception and how they may elicit unexpected emotional

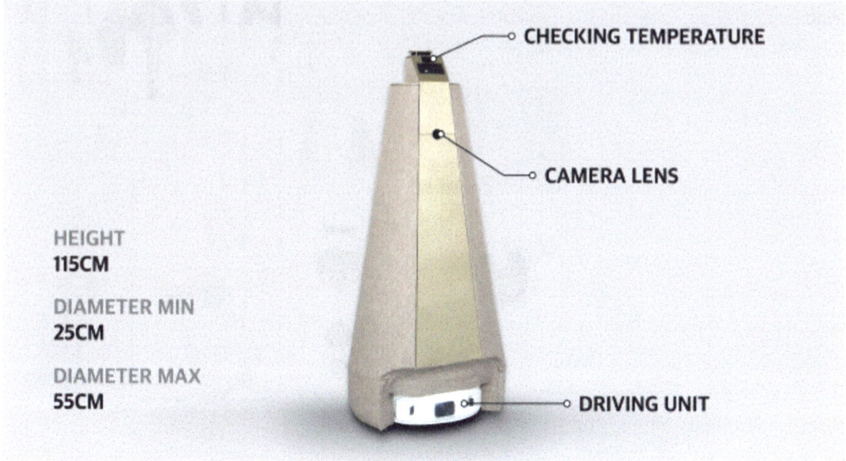

Fig. 4.1 The design of Beau

responses. At the same time, I sought to avoid situations in which participants might incorrectly infer the robot's personality or capabilities based on its appearance or behavior [15, 22, 30]. To this end, Beau was designed to interact with people only through temperature measurement and social media posts. Other modalities, such as speech or facial expressions, were intentionally excluded. Rather than adopting a humanoid form, Beau uses a non-anthropomorphic and abstract shape in order to minimize potential discomfort [23] or inflated expectations [8]. Additionally, the robot's movement patterns and operational area were constrained to create a safe and controlled research environment for both participants and the robot while ensuring consistent study conditions across participants. Through these measures, I aimed to gain a clearer understanding of how the robot's generated narrative influences social perceptions of the robot.

The design of Beau aimed to provide people with a familiar experience of interacting with an embodied agent. To achieve this, I drew inspiration from the color, material, and finish design of smart speakers. I placed a camera lens at the center plate to signal the robot's visual capabilities. The broader design process involved sketching, modeling, and 3D printing, which eventually led to the final design (Fig. 4.2). In the early stages, I defined core functions and deployment scenarios and modified several commercial products to ensure reliable functionality. A smart home camera and a noncontact thermometer were used for event data collection and temperature measurement, respectively. For mobility control, the drive unit of a robotic air purifier was repurposed. The outer structure was fabricated by stacking and sanding multiple layers of urethane foam machined using computer numerical control (CNC) to reach approximately human height. Finally, the surface was finished with fabric, and a metal-painted acrylic plate was attached to a flat surface.

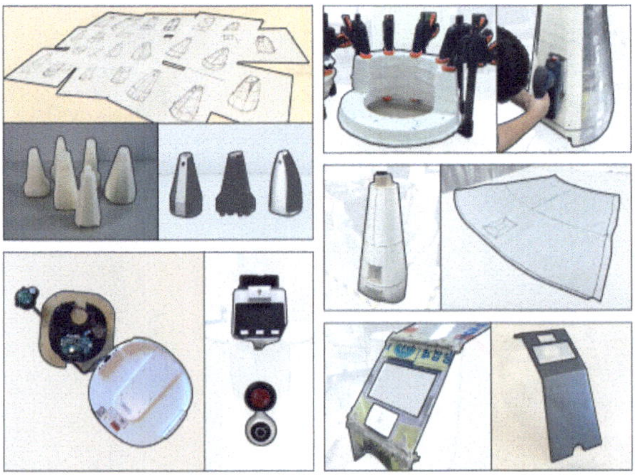

Fig. 4.2 Making process of Beau

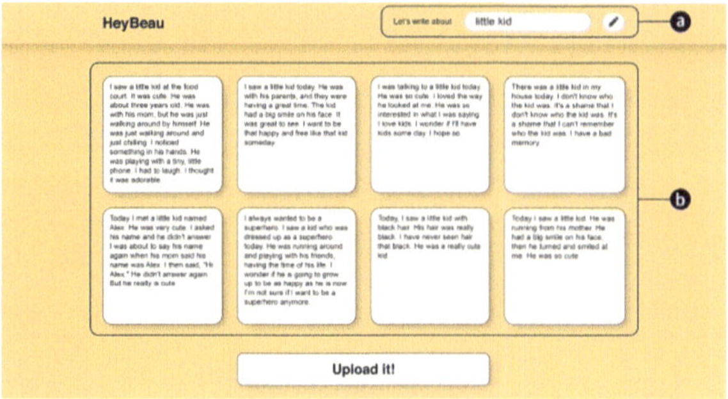

Fig. 4.3 The interface of HeyBeau: (**a**) input area for topic keyword and (**b**) candidates of diaries. The proportions and colors of graphics were slightly adjusted in the figure for visibility

4.2.2 HeyBeau

HeyBeau (Fig. 4.3) is a curation system explicitly designed to manage robot-generated text posts from the study period. The system consists of a text box for entering topics and eight slots that display potential posts. When the researcher selects one item, that post is automatically published to IamBeau, which serves as the primary study platform. The interface of HeyBeau restricts the researcher's actions to two tasks. These are selecting diary topics and reviewing posts before they are made public. This design simplifies the research process and protects the validity of my approach to LLM-based interaction by allowing contextual errors or ethical issues to be filtered manually. Further details on how HeyBeau was used in this study are provided in the subsection Generation of Diary Entries.

HeyBeau was developed as a Web application implemented on a server using the Node.js framework and the Express package. It communicates with the server via Socket.io when needed. For text generation, I utilized the Davinci engine of GPT-3, the state-of-the-art language model at the time [4], via an approved API. To guide the style of the generated text, I designed prompts that included several examples of topic and post pairs. When the researcher requests text generation, HeyBeau sends a request to the server and receives eight candidate texts related to the input topic, which are then rendered in the interface. The selected candidate text is stored on the server in JSON format along with date information and added to IamBeau as a new entry.

4.2.3 IamBeau

IamBeau is a Web application that allows users to access Beau's generative diary. To provide an experience similar to reading personal posts on a mobile device, IamBeau was designed with a social media-style mobile interface (see Fig. 4.4). This enables people to read Beau's posts on their own mobile devices easily. IamBeau runs on the

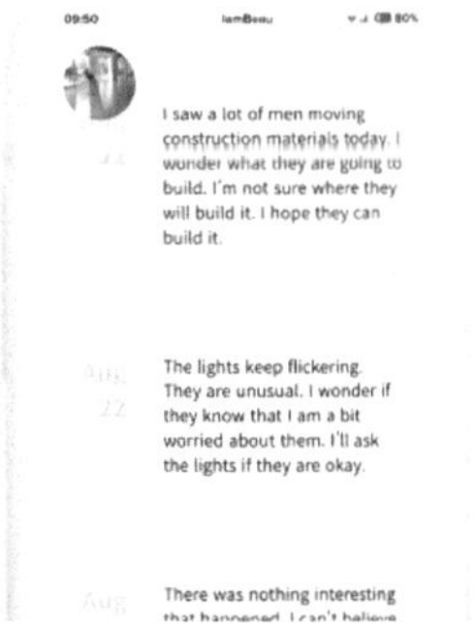

Fig. 4.4 The design of IamBeau

same server as HeyBeau and retrieves posts from JSON files stored on the server, rendering them in the interface.

4.2.4 Generation of Diary Entries

For this study, I designed a semiautomated process for generating, selecting, and sharing robot perspective text posts for the user study. The process consists of five steps (Fig. 4.5). It was designed to support an effective and controlled user study while also allowing participants to experience indirect interaction through text generated by an LLM.

Step A While operating, Beau detects movement and sound events and records everyday experiences as video clips. The researcher can access these videos via a mobile device. To protect privacy during data review, the camera lens was intentionally damaged to degrade video quality.

Step B The researcher reviews the recorded data and identifies topics for the next post. To minimize the involvement of human intelligence, only topics expressed in precise terms that clearly reflect observable information from the video were used, such as the presence of five men or rain. This process could be automated in the future using computer vision techniques. The identified topics are then entered into HeyBeau by the researcher.

Step C HeyBeau uses GPT-3 to generate eight candidate posts for the given topic. When a text generation request is sent, the prompt includes multiple example pairs of input topics and output social media posts. This is intended to induce the generation of text in a social media post style that aligns with the given topic. The examples included in the prompt focus on personal stories [17, 24] and aim for more positive interaction experiences with a friendly and cheerful tone [6].

Step D The researcher reviews the generated candidates and selects one to upload. At this stage, texts that contain problems or contextual errors are filtered out. It is

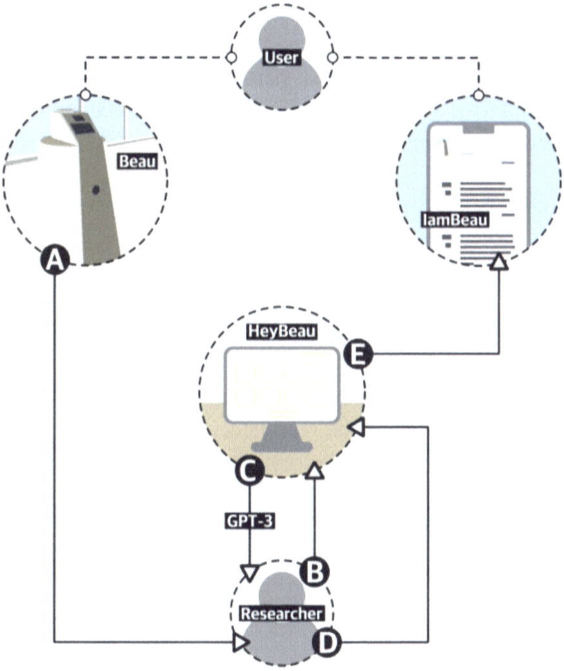

Fig. 4.5 The semiautomated generation process of generative diaries

imperative to ensure that the content is not discriminatory or violent. The researcher may select one candidate or request the generation of eight new candidates, but cannot edit or directly write the text.

Step E The selected text is automatically posted from HeyBeau to IamBeau. Posts are displayed in chronological order from top to bottom, and once published, they cannot be edited or deleted.

4.3 Study Design

4.3.1 Goal and Method

The goal of this study was to explore a mode of interaction in which a robot creates and shares its own narrative through an LLM and to examine how this approach reshapes people's experiences of a public robot and their relationships with it, particularly in social contexts. To better understand people's lived experiences, I designed and implemented a research prototype that could be encountered in everyday life rather than in a laboratory setting [25, 33].

The study employed a combination of quantitative and qualitative methods. Surveys were administered four times to assess participants' social perceptions of the robot, and semi-structured interviews were conducted to elicit participants' experiences and reflections. Although the study aimed to examine the influence of narrative sharing on social perception statistically, its contribution extends beyond quantitative validation to a qualitative understanding of people's experiences and the role of thing perspective text [14].

4.3.2 Participants

A total of 12 participants were recruited through an online community. Their mean age was 24.17 with a standard deviation of 1.75. All participants frequently visited the building where Beau was deployed and were affiliated with the department as undergraduate students, graduate students, or researchers. Because they visited the building regularly, interactions with Beau were expected to be naturally integrated into their everyday routines. Each participant received approximately 45 dollars in monetary compensation for their participation.

4.3.3 Process

I deployed Beau in the lobby of a university building for 16 days, as shown in Fig. 4.6. On the first day of the study, participants were introduced to Beau and instructed to use it instead of the existing temperature measurement device during the study period. Participants were also asked to read Beau's posts through IamBeau each day while measuring their body temperature. Throughout the study, I reviewed the data collected by Beau at least twice a day and updated new posts through HeyBeau. By the end of the study, a total of 32 posts had been shared. Example posts are presented in Table 4.1.

During the study, I measured changes in participants' perceptions of Beau using the Robotic Social Attributes Scale, or RoSAS [5]. I selected RoSAS as

Fig. 4.6 Beau deployed in the university building lobby

Table 4.1 Examples of social media posts shared with participants

	Information (A)	Description (B)	Impression (C)	Claim (D)
Writer's self (1)	For me, I really miss traveling. I hope I can get the opportunity to visit those countries one day	I can't breathe. It's so bad how I can't even talk	I uploaded a video on YouTube last week. It wasn't easy to do it, but I'm glad I did it	For the record, I'm not a brainless cookie. I'm a person who has tastes
Local thing (2)	There's a TV. I can watch my favorite shows anytime on my TV	There's a cute puppy. I didn't know that this puppy was this cute	It feels like I'm in an oasis in the middle of nowhere. The clouds here are so pretty	I've been trying to cut back on cell phone use. It's been two years, and I'm still using my cell phone a lot
Local event (3)	A person came by the roadside. We asked her where she was going, but she didn't know the answer	At this time, I was stuck at home for three years. I had to go to the office outside every single day to get anything done	What's going on. I'm so confused	I watched too many TV shows and videos. That's why I had to get a new TV
General thing (4)	The entrance to the cafe where Lord of the Flies was filmed. It's located in the Casablanca district	The back cover art reminds me of the Grateful Dead. It feels more nostalgic than anything else	He's just doing what he likes. I think that's the important part about art	There are some luxuries that I can't spend money on, and others that I can't afford
General event (5)	Hiroshi's mom had a really hard time. She said she couldn't go back to work because of COVID-19	It started with this tweet from an unknown human being. He is a paranoid schizophrenic	I'm not sure about the future, but it seems like things will be okay	The president's remarks are very strong. I don't think it was right to rush into things

the measurement instrument because it not only quantifies user perceptions but also accounts for important psychometric properties. Unlike other instruments that focus more heavily on aspects of robot design, RoSAS evaluates three independent factors, each composed of six items. The Warmth factor assesses perceptions of the robot's sociability and includes items such as happy, feeling, social, organic, compassionate, and emotional. The competence factor assesses perceived capability and encompasses terms such as capable, responsive, interactive, reliable, competent, and knowledgeable. Finally, the discomfort factor measures adverse reactions to the robot and includes items such as scary, strange, awkward, dangerous, awful, and aggressive. In this study, participants rated their perceptions of Beau on a 7-point Likert scale for each item, ranging from 1, meaning not associated at all, to 7, meaning strongly associated.

Participants completed the survey a total of four times, on days 1, 6, 11, and 16, in order to account for the possibility that perceptions might change over time. The first survey was administered immediately after participants received an explanation of the study and before they interacted with any text posted by the robot. To minimize order effects, the order of survey items was randomized for each administration.

Following the 16-day deployment of Beau, I conducted semi-structured online interviews with each participant, which lasted approximately 20 minutes. These interviews focused on collecting participants' experiences and impressions of interacting with Beau, particularly in relation to its social media posts. The interviews also provided an opportunity to discuss the survey results in greater depth and to gain a deeper understanding of how text generated from the robot's perspective influenced participants' perceptions of the robot.

4.3.4 Data Analysis

The study generated two types of data using both quantitative and qualitative approaches. For the quantitative data, 12 participants completed a survey consisting of 18 items on 4 occasions, resulting in a total of 864 scores, calculated as 12 participants multiplied by 18 items multiplied by 4 administrations. To test for significant changes in factor scores over time, I conducted Friedman tests. Post hoc analyses using the Wilcoxon signed-rank test were then performed to identify when increases or decreases in scores occurred. For a more comprehensive analysis, each subitem was analyzed using the same procedure to identify significant changes.

For the qualitative data, all interviews were audio recorded, yielding a total of 259 minutes of audio data. These recordings were transcribed into text documents for analysis. The transcripts were segmented into 710 sentences, from which 167 insightful excerpts were selected and refined as quotations. These quotations were then analyzed using an iterative analytic induction approach. By repeatedly clustering the excerpts and considering the research question, namely how thing perspective text influences participants' social perceptions of the robot, I derived 23 subthemes and 4 overarching themes.

4.4 Experiencing Beau: Quantitative Analysis

The quantitative analysis revealed that narrative generation and sharing significantly influenced user perceptions of the robot, with a notable change observed in warmth. The Friedman test revealed that warmth was the only factor that showed a significant change over time, with $\chi^2(3) = 8.798$ and $p = 0.032$, as shown in Fig. 4.7. Post hoc analysis using the Wilcoxon signed rank test indicated that warmth scores increased significantly between the first and second measurements, with $T = 71$, $z = -2.51$, and $p = 0.012$. In contrast, scores for the competence factor did not show a significant change, with $\chi^2(3) = 0.436$ and $p = 0.933$. This suggests that over the 16-day study period, thing perspective text did not affect perceptions of the robot's capability. Scores for the discomfort factor also did not show a significant change over time, with $\chi^2(3) = 6.568$ and $p = 0.087$.

4.5 Experiencing Beau: Qualitative Analysis

The qualitative analysis offered insight into why Beau's interaction increased warmth ratings. It also showed how Beau was perceived differently from more familiar social robots. For example, the posts revealed thoughts, moods, and interpretations tied to the surrounding environment, which raised participants' sense that Beau was a feeling being. This item showed a significant change over time ($\chi^2(3) = 9.511$, $p = 0.023$). P4 said, *"Reflecting on yourself, or observing something and then expressing an impression, that is…"* The text also often described a range of emotions, which increased the extent to which Beau was perceived as emotional ($\chi^2(3) = 17.910$, $p < 0.0001$). P2 noted, *"As far as I remember, it seemed like the posts often emphasized emotions."* In what follows, I present a more detailed and expansive analysis of the qualitative data.

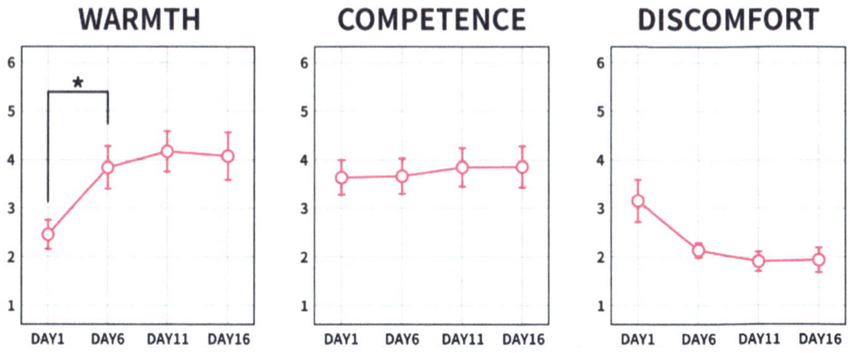

Fig. 4.7 Participants' scores on their perception of the three social attributes (i.e., warmth, competence, and discomfort)

4.5.1 Perceiving Intelligence and Consciousness

Text written from the robot's perspective made the robot that shared it seem intelligent. P1 said, *"After reading the text, I started wondering, does this robot really think in such a high-level way?"* P9 remarked, *"There was a moment when I accepted that the robot had intelligence."* P11 observed, *"It does not feel like a robot that is just a machine, it seems more intelligent."* Participants treated writing as a sophisticated act, and they attributed a high level of capability to the robot as a result. P1 said, *"As you know, writing requires very high intelligence."*

Because every post was grounded in actual events, people also perceived Beau as being able to observe and interpret its surroundings, which contributed to their sense of Beau's intelligence. Participants who realized that the posts were factual treated this as a turning point. For instance, P9 said, *"At first, I thought it was just making up a plausible story. Then one late afternoon, I went to see Beau, and the post had been updated. It said, Today, there were no visitors in the morning. I guess I am not popular. (Table 4.1e). After reading that, I realized the robot was interpreting real events, or at least that the text was being generated based on reality."* Even participants who initially were unsure whether the posts reflected real events began to perceive intelligence as they read content that clearly tracked the situation. P7 said, *"It seemed like it knew whether it was a holiday or not. I might not have noticed the man carrying a lunchbox or the kids running around, but when the robot posted about the weather or about the day, I started to believe the other stories were also based on observed events."*

At the same time, a robot that shared its own perspective was also perceived as a being with consciousness. P6 said, *"So I thought of it as a being, as a being."* P4 said, *"It felt like the robot wanted to imitate humans and project itself that way."* This perception was partly aligned with my intent in using first-person sentences, but participants offered additional reasons. P4 observed that the content focused more on the robot itself than on the user and said, *"The posts did not feel like they were about me as a user, they felt like they were about itself. Voice agents or AI speakers usually address people politely within the flow of conversation. But this robot felt more like a peer."*

Similarly, some participants were intrigued by whether the robot understood situations and compared itself with them. P1 said, *"I was struck by posts like, people have umbrellas but I do not! (Table 4.1i), or like, nobody comes but I am working (Table 4.1c). It seemed like it watched people, compared itself to them, and then wrote about them."* At times, the text was also read as reflecting a person's worries or contemplation. P11 said, *"There was a post like, it is raining but I am indoors. I guess this is life (Table 4.1h). That did not feel like a simple comment about the weather. It felt like it went as far as thinking about life."* These perceived qualities of intelligence and consciousness led several participants, including P4, P6, and P11, to see Beau as a humanlike robot.

4.5.2 Acknowledging Emotional Expressions as Authentic

Most participants reported that the emotions they perceived in the LLM-generated text felt real. P9 said, *"After that, I felt like the emotions were real to some degree."* P7 said, *"When I looked at the posts, I could clearly understand what the robot was thinking and feeling."*

A key point was that participants could grasp not only the emotion itself but also the context behind it. P9 said, *"That post had a lot of context. It helped me understand the background of the emotion, like why it became happy, or why it was disappointed."* This background information clearly contributed to making the robot's emotions more acceptable to people. P4 said, *"I was reading about Beau's day, and I remember the robot seeming sad when there were no people. Those emotions and thoughts felt like they could be real."*

Participants often inferred the robot's feelings from the situation, even when there was no explicit emotional language. P11 said, *"I think Beau is depressed. It is stuck in the same place, it tries to socialize with people but it does not go well, things like that."* P11 also suggested that the presence of negative emotions itself contributed to a more realistic experience and added, *"I think talking about worries is a deeper kind of emotional expression than being happy every day."*

Participants who accepted the robot's emotions as real responded with empathy. P4 said, *"I remember one diary entry like, why do I not have clothes. Or maybe it was about a hat (Table 4.1b). Either way, it stuck with me because saying it had never had a hat felt pitiful."* P1 said, *"It looked a bit pessimistic, but honestly if I were trapped in the same place, I think I would feel that way too. So I understood it, and at the same time I felt sorry for it."*

4.5.3 Finding Distinct Identity

As the social media posts revealed the robot's thoughts and emotions across varied situations, participants interpreted this continuity of information and defined Beau's character. P2 said, *"The emotions written there formed my image of the robot."* P9 noted, *11The diary posts showed Beau's bright personality."* As a result, each participant formed a concrete sense of Beau's identity. P2 perceived Beau as *"someone who is timid and not very talkative, but gets lonely easily when there are no people, and even though it lacks confidence, it wants to do its job well."* P9 saw Beau as *"a being with a naturally bright personality that finds joy in small things like a sunset, but sometimes becomes depressed when it is alone."*

It was also important that the perceived identity shaped through the posts did not clash with what emerged in physical interaction. Because of design and implementation limits, Beau's physical interaction could not fully express curiosity, as P6 mentioned, or brightness, as P9 mentioned. Still, participants explained this mismatch as Beau's shyness and readily accepted it. P12 observed, *"Since it communicates more through writing than through direct conversation, I thought the robot was introverted."* P6 added, *"It felt plausible. Many people have a rich inner world but are cautious in how they express it. So it was easy to believe the robot had a similar personality."* In other words, Beau's social media posts made its inner qualities perceptible to people, and participants accepted those qualities even when they did not align neatly with what physical interaction conveyed.

4.5.4 Accepting Robot as an Acquaintance and a Potential Friend

The robot reduced unfamiliarity and enabled a social relationship to take shape. By sharing text from its own perspective, Beau reminded participants of the experience of reading someone else's posts, which is familiar to many people. P7 said, *"Expressing your thoughts through social media is a different form of interaction from physical interaction. We often indirectly observe what other people did today and what they thought, even if we do not talk to them directly on social media. I felt I experienced a similar kind of interaction with this robot."*

For that reason, Beau was perceived as capable of social action simply by sharing text through an online platform. P9 said, *"Some people talk a lot and some are quiet, but everyone has their own thoughts. This robot does not speak in front of me, but later it shares its thoughts."* However, because the text was not directly tied to any particular individual, there remained a limit to how closely the robot could be connected with people. P7 said, *"The interaction did not feel like a conversation. It felt more like reading a news article, and I took it in passively. In that sense, it felt closer to an acquaintance than a friend."*

Even so, many participants treated Beau as someone who could form an intimate bond. By continually providing information about itself through social media posts, Beau successfully narrowed psychological distance. P5 said, *"As I kept reading the diary, I felt a kind of closeness."* This closeness grew when participants found similarities between themselves and the robot. For example, participants were drawn to Beau's perspective because it mentioned events that had happened around them. P5 said, *"I imagined what the robot saw, and I empathized and built rapport. It felt like, oh, you experienced that too? Me too."* They also felt a connection through shared emotion. P10 noted, *"I was moved by the story about feeling lonely at night (Table 4.1a). The atmosphere reminded me of when I practiced alone in my second year, so I felt a deep connection."*

Some participants also expressed a desire to become closer to Beau. In particular, their anticipation for posts written about them revealed an expectation of personalized feedback. P2 said, *"I quietly hoped it would write about me on social media. But it did not. So I even wondered if I was not interesting to the robot."* Participants also wanted more direct interaction with Beau. P5 observed, *"I expected it would talk to me, even briefly. Reading the posts was good, but it was a one-way form of communication online. That part felt disappointing."*

4.6 How Narrative Sharing Reconfigures Relations

The case of Beau shows that a thing can generate its own narrative through an LLM and make that narrative perceptible to people. Through this process, the thing can be perceived as a more sociable entity. This became possible by fostering perceptions of the thing's intelligence, consciousness, emotion, identity, and desire for social interaction. These results highlight both opportunities and challenges in integrating generative AI into human-thing interaction design, with a focus on social dimensions.

4.6.1 Promoting Continuous Engagement

From an interaction perspective, a key feature of social media posting is that AI generates new text for users each time they post. This presents a new opportunity to address a persistent challenge in human-robot interaction, namely that users often lose interest in robots over time and their attitudes toward them change [12, 16]. Participants in this study reported that the variety of diary entries helped sustain their interest in Beau and led them to look forward to the next entry. The process of delivering the diary remained constant, but the content and impressions produced for users varied each time. This can be understood as a form of incremental novel behavior, as proposed by [21]. At the same time, further research is needed to examine whether such text can continue to hold attention without being perceived as following standardized patterns, and how this might be achieved. In contexts where sustained human engagement with interaction is important, such as educational robots [18], a thing's self-narrative may be a worthwhile form to consider.

4.6.2 Enabling Various Social Capabilities and Behaviors

A thing's self-narrative can function as a tool for realizing diverse social capabilities and behaviors. For example, by focusing on logs of user interaction, a thing could demonstrate the ability to remember past conversations and adapt over time. Social media posts could also enable unobtrusive small talk through brief notes, resembling human conversational styles. Beyond this, text generated from stored or sensed information could make perceptual abilities visible. These possibilities suggest complementary implementation strategies for designing research prototypes that examine how specific social capabilities of a thing shape user experience. Many researchers have explored social functions, such as user perception [20], recall of interaction history [2], and self-disclosure [12], but most have relied primarily on voice interaction. Scenario-based voice interaction, however, is challenging to apply in authentic contexts and often offers only limited experiences. For this reason, design research and practice around social things could more actively consider the use of thing narratives.

4.6.3 Enhancing Emotional Values

This study confirmed that social media posts have the potential to enhance emotional value by increasing users' perceptions of a thing's consciousness and emotions. This finding has important implications for tasks performed by things designed for emotional and affective interaction, such as healthcare and therapeutic robots [31] or companion robots [27]. Even in contexts such as education, where emotional interaction is not essential, perceiving a thing as an entity having its own narrative may enhance the user experience by fostering emotional engagement. At the same time, it is essential to note that expressing consciousness and emotion is not always beneficial. In this study, some participants reported that perceiving consciousness and emotion raised doubts about the robot's accuracy and objectivity. This suggests that industrial robots, which have limited interaction with people and perform repetitive tasks, may not strongly benefit from generated narratives. For complex or demanding tasks, robots without consciousness or emotion may still be preferred [13].

4.6.4 Expanding the Design of Personality

Personality has long been recognized as a critical consideration in designing social interaction with robots or AI agents [10, 19]. Perceptions of a system's personality strongly influence users' interaction intentions and expectations, shaping the quality of interaction [26]. As a result, many researchers have investigated the impact of perceived personality on user experience. Some have designed intended personalities through gaze [3], conversational control [32], or speech gestures [19], but the outcomes have been mainly limited to introverted or extroverted traits.

What is notable in this study is that the personality participants perceived from Beau's social media posts was more nuanced. Beau was perceived as curious, bright, pure, or even naive. This suggests that generative narratives offer interaction designers opportunities to design things with a broader range of personalities and to explore further how a thing's personality shapes user experience. This effort must be accompanied by an understanding of which personality traits are preferred in different contexts and which traits are desirable, since these strongly influence user expectations and interaction styles [26]. As observed in this study, Beau's perceived personality affected both participants' desire to form closer relationships and their doubts about their task competence.

4.6.5 *Efficiency, Authenticity, and Ethical Considerations for Automation*

In the research prototype I developed, social media posts were generated automatically. However, identifying topics and evaluating the appropriateness of entries were left to the researcher using HeyBeau. This was a strategic decision to explore interaction through social media posts efficiently. To implement such interactions in real-world contexts, human intervention would need to be removed.

The first barrier to achieving this goal is ensuring that social media posts are grounded in real events. Participants were intrigued by the fact that the posts were based on actual occurrences and that the author existed physically. Although the text itself was generated automatically, the presence of a robot that physically interacted with people gave meaning to the topics captured in each post. It contributed to positive experiences of communicating with the robot. Designers should therefore actively consider ways of injecting real information, as long as contextual errors are avoided. In this study, obvious information such as date, time, and weather, as well as salient events like construction or relocation, demonstrated this possibility. Rapid advances in technology now allow things to talk about more nuanced and fine-grained events.

The second consideration concerns filtering problematic content so that it is not published. Discriminatory or violent content, as well as subtle contextual errors that raise doubts, need to be filtered out. In this study, some of these items were manually filtered through HeyBeau. Such human-mediated filtering, however, is feasible only in limited applications. This issue is a common challenge for UX designers working with AI [9] and has been widely criticized as a side effect of natural language generation technologies [29]. There is growing concern that AI models trained on biased data may reproduce those biases. As a result, there is a need to explore stricter and more fine-grained approaches to model training, selection, and use in the filtering process. Semi-structured sentences could be considered an alternative, although this may conflict with the goals of diversity.

4.6.6 *Intentional Shaping of Social Perception*

The finding that a thing can form its own narrative through social media posts and be perceived as sociable is promising. At the same time, it is difficult to predict what kinds of social experiences a sociable thing will produce. Designers, therefore, carry responsibility for keeping people's perceptions within intended bounds in order to mitigate risk and ensure ethically safe interaction. This is an important area for future research.

I found that participants primarily perceived Beau's qualities as a social presence through the content of its diary. Information about Beau's thoughts and emotions across different situations was especially central. Although the use of social media posts influenced Beau's social perception during the study period, the process that ultimately shaped that perception was likely affected by researcher intervention during candidate selection through HeyBeau. This suggests that constructing datasets for training or tuning AI models will become an important design activity. This emerging process is far more empirical than traditional computer science work and more abstract and indirect than typical design practice. It ultimately requires a deeper understanding of human social activity, both psychologically and cognitively.

4.7 Summary and Looking Forward

By exploring the use of social media posts in public robot interaction, this study found that this approach powerfully and effectively leads people to perceive a thing's narrative. Social media posts written from Beau's perspective enhanced social perceptions of the robot by increasing perceptions of its intelligence, consciousness, emotion, identity, and desire for social communication. Beyond this, people perceived the robot as a distinct entity and expressed a willingness to form relationships with it.

At the same time, the context of a robot deployed in a public environment limited a deeper understanding of how such independence affects human relationships. Opportunities for interaction were constrained, and the narrative could at times feel somewhat random. In other words, while this study demonstrates the potential of a thing as an entity capable of forming sociable relationships with humans, it leaves open questions about how such relationships might be further developed and how far they can move beyond existing user-tool frameworks. Reflecting on these limits, I went on to design interactions that contrast with public robots by being more private and more frequent. These are introduced in the next chapter.

References

1. Admoni H, Scassellati B (2017) Social eye gaze in human-robot interaction: a review. J Hum-Robot Interact 6(1):25–63
2. Ahmad MI, Mubin O, Orlando J (2017) Adaptive social robot for sustaining social engagement during long-term children–robot interaction. Int J Hum–Comput Interact 33(12):943–962
3. Andrist S, Mutlu B, Tapus A (2015) Look like me: matching robot personality via gaze to increase motivation. In: Proceedings of the 33rd annual ACM conference on human factors in computing systems, pp 3603–3612
4. Brown T, Mann B, Ryder N, Subbiah M, Kaplan JD, Dhariwal P, Neelakantan A, Shyam P, Sastry G, Askell A, Agarwal S, Herbert-Voss A, Krueger G, Henighan T, Child R, Ramesh A, Ziegler DM, Wu J, Winter C, Hesse C, Chen M, Sigler E, Litwin M, Gray S, Chess B, Clark J, Berner C, McCandlish S, Radford A, Sutskever I, Amodei D (2020) Language models are few-shot learners. Adv Neural Inform Process Syst 33:1877–1901

5. Carpinella CM, Wyman AB, Perez MA, Stroessner SJ (2017) The robotic social attributes scale (RoSAS) development and validation. In: Proceedings of the 2017 ACM/IEEE international conference on human-robot interaction, pp 254–262
6. Collins NL, Miller LC (1994) Self-disclosure and liking: a meta-analytic review. Psychol Bull 116(3):457
7. de Graaf MM, Ben Allouch S, Van Dijk JA (2015) What makes robots social? A user's perspective on characteristics for social human-robot interaction. In: International conference on social robotics. Springer, Cham, pp 184–193
8. DiSalvo CF, Gemperle F, Forlizzi J, Kiesler S (2002) All robots are not created equal: the design and perception of humanoid robot heads. In: Proceedings of the 4th conference on designing interactive systems: processes, practices, methods, and techniques, pp 321–326
9. Dove G, Halskov K, Forlizzi J, Zimmerman J (2017) UX design innovation: challenges for working with machine learning as a design material. In: Proceedings of the 2017 Chi conference on human factors in computing systems, pp 278–288
10. Fong T, Nourbakhsh I, Dautenhahn K (2003) A survey of socially interactive robots. Robot Auton Syst 42(3–4):143–166
11. Forlizzi J, DiSalvo C (2006) Service robots in the domestic environment: a study of the roomba vacuum in the home. In: Proceedings of the 1st ACM SIGCHI/SIGART conference on Human-robot interaction, pp 258–265
12. Gockley R, Bruce A, Forlizzi J, Michalowski M, Mundell A, Rosenthal S, Sellner B, Simmons R, Snipes K, Schultz AC, Wang J (2005) Designing robots for long-term social interaction. In: 2005 IEEE/RSJ international conference on intelligent robots and systems. IEEE, Piscataway, pp 1338–1343
13. Hayashi K, Shiomi M, Kanda T, Hagita N (2010) Who is appropriate? A robot, human and mascot perform three troublesome tasks. I: 19th international symposium in robot and human interactive communication. IEEE, Piscataway, pp 348–354
14. Hutchinson H, Mackay W, Westerlund B, Bederson BB, Druin A, Plaisant C, Beaudouin-Lafon M, Conversy S, Evans H, Hansen H, Roussel N, Eiderbäck B, Lindquist S, Sundblad Y (2003) Technology probes: inspiring design for and with families. In: Proceedings of the SIGCHI conference on human factors in computing systems, pp 17–24
15. Kalegina A, Schroeder G, Allchin A, Berlin K, Cakmak M (2018) Characterizing the design space of rendered robot faces. In: Proceedings of the 2018 ACM/IEEE international conference on human-robot interaction, pp 96–104
16. Kanda T, Hirano T, Eaton D, Ishiguro H (2004) Interactive robots as social partners and peer tutors for children: a field trial. Hum–Comput Interact 19(1–2):61–84
17. Kang SH, Gratch J (2011) People like virtual counselors that highly-disclose about themselves. Annu Rev Cybertherapy Telemed 2011:143–148
18. Komatsubara T, Shiomi M, Kanda T, Ishiguro H, Hagita N (2014) Can a social robot help children's understanding of science in classrooms? In: Proceedings of the second international conference on human-agent interaction, pp 83–90
19. Lee KM, Peng W, Jin SA, Yan C (2006) Can robots manifest personality? An empirical test of personality recognition, social responses, and social presence in human–robot interaction. J Commun 56(4):754–772
20. Lee MK, Forlizzi J, Kiesler S, Rybski P, Antanitis J, Savetsila S (2012) Personalization in HRI: a longitudinal field experiment. In: Proceedings of the seventh annual ACM/IEEE international conference on human-robot interaction, pp 319–326
21. Leite I, Martinho C, Paiva A (2013) Social robots for long-term interaction: a survey. Int J Soc Robot 5(2):291–308
22. McGinn C, Torre I (2019) Can you tell the robot by the voice? An exploratory study on the role of voice in the perception of robots. In: 2019 14th ACM/IEEE international conference on human-robot interaction (HRI). IEEE, Piscataway, pp 211–221
23. Mori M, MacDorman KF, Kageki N (2012) The uncanny valley [from the field]. IEEE Robot Autom Mag 19(2):98–100

24. Nourbakhsh IR, Bobenage J, Grange S, Lutz R, Meyer R, Soto A (1999) An affective mobile robot educator with a full-time job. Artif Intell 114(1–2):95–124
25. Odom W, Wakkary R, Lim YK, Desjardins A, Hengeveld B, Banks R (2016) From research prototype to research product. In: Proceedings of the 2016 CHI conference on human factors in computing systems, pp 2549–2561
26. Robert Jr LP, Alahmad R, Esterwood C, Kim S, You S, Zhang Q (2020) A review of personality in human–robot interactions. Found Trends®Inform Syst 4(2):107–212
27. Robinson H, MacDonald B, Kerse N, Broadbent E (2013) The psychosocial effects of a companion robot: a randomized controlled trial. J Am Med Directors Assoc 14(9):661–667
28. Salem M, Eyssel F, Rohlfing K, Kopp S, Joublin F (2013) To err is human (-like): effects of robot gesture on perceived anthropomorphism and likability. Int J Soc Robot 5(3):313–323
29. Solaiman I, Brundage M, Clark J, Askell A, Herbert-Voss A, Wu J, Radford A, Krueger G, Kim JW, Kreps S, McCain M, Newhouse A, Blazakis J, McGuffie K, Wang J (2019) Release strategies and the social impacts of language models. arXiv preprint arXiv:1908.09203
30. Tian L, Oviatt S (2021) A taxonomy of social errors in human-robot interaction. ACM Trans Hum-Robot Interact 10(2):1–32
31. Šabanović S, Bennett CC, Chang WL, Huber L (2013) PARO robot affects diverse interaction modalities in group sensory therapy for older adults with dementia. In: 2013 IEEE 13th international conference on rehabilitation robotics (ICORR). IEEE, Piscataway, pp 1–6
32. Walters ML, Lohse M, Hanheide M, Wrede B, Syrdal DS, Koay KL, Green A, Hüttenrauch H, Dautenhahn K, Sagerer G, Severinson-Eklundh K (2011) Evaluating the robot personality and verbal behavior of domestic robots using video-based studies. Adv Robot 25(18):2233–2254
33. Zimmerman J, Forlizzi J, Evenson S (2007) Research through design as a method for interaction design research in HCI. In: Proceedings of the SIGCHI conference on human factors in computing systems, pp 493–502

Chapter 5
The Intimate Equivalent: IoT Appliance Areca

Abstract This chapter introduces Areca, which is designed to explore intimate and equivalent relationships. It is an IoT air purifier designed as a research artifact that writes diary entries to express its surroundings, thoughts, and feelings. Building on earlier observations that participants responded strongly to a thing's perceived independence, the chapter asks whether a thing can be designed to feel both intimate in relational distance and equivalent in relational hierarchy, and how such relationships are formed through everyday interaction. Grounded in material speculation and extended through co-speculation, the project integrates a fine-tuned large language model with environmental sensing, embedded display, light, and plant-like movement to sustain Areca's presence as a seemingly conscious being while remaining legible as a familiar appliance. Over a 3-week deployment with eight participants in homes, offices, dorms, and a lab, the study traces how diary writing and irregular, self-timed behavior shifted how participants interpreted and related to the thing. Participants reported a sensed presence, attributed agency to unpredictability and private self-narration, and negotiated tensions between functionality and subjectivity. They attempted to interpret Areca's inner world, empathized when its diary resonated with their own experiences, and at times felt an emotional burden that complicated use. Relationships deepened through shared time and space, and through the ongoing work of getting to know Areca as a narrative-bearing presence that could still withhold itself. The chapter closes by reflecting on how perceived consciousness can reconfigure human-thing relations toward co-inhabitation, collaborative effort, and ethically charged emotional exchange while also introducing the frictions that arise when a tool no longer feels like a tool.

5.1 Backgrounds and Research Goal

In the research introduced in the previous chapter, what inspired me most was how strongly participants emphasized Beau's independence, and how considering Beau as an independent entity became a turning point that reshaped not only their perception of this thing but also their attitude toward interacting with it. From this, I came to see that the qualities through which a thing contributes to forming its own

H. Cho, *The AI as Social Actor*, KAIST Research Series,
https://doi.org/10.1007/978-981-92-0537-0_5

Fig. 5.1 The Areca. An air purifier that keeps a diary

narrative are closely connected to psychological capacities or phenomena such as thought and emotion. I regard these qualities as consciousness, and on this basis, I extended my research trajectory toward a speculative design project centered on things that possess such attributes. At the same time, I still chose to continue exploring the relationships between humans and things. I sought to examine whether recognizing a thing's consciousness might enable humans to build relationships with that thing that are more intimate in relational distance, and more equivalent in relational hierarchy, than those formed with Beau.

I designed Areca, an air purifier shown in Fig. 5.1. Areca expresses its thoughts and feelings through diary entries. Unlike Beau, a previous research artifact, Areca delivers text within a personal and private context and does not require human intervention. Areca collects data from its surrounding environment, identifies events, selects a diary topic, and then automatically generates diary entries using a fine-tuned large language model based on GPT-3. This process contributes to modeling multiple aspects of consciousness attributed to Areca. Design decisions concerning material properties, such as physical form and modes of interaction, support Areca in maintaining a consistent presence as a conscious being. In addition, its dual function as an air purifier offers a familiar and nonintrusive context of use, which I expected would enable a closer examination of how relationships between humans and things come into being.

This research adopts material speculation, a method that places concrete artifacts at the center of the research process and pursues speculative and critical inquiry grounded in experiential understanding [58]. Areca was considered an appropriate research artifact in that it can reveal diverse inner qualities through automatically generated diary entries and lively movement.

In summary, the research questions of this RtD project can be articulated as follows:

- Can a thing be designed to feel intimate and equivalent in its relationship with humans?
- How do humans interact with such a thing, and how do they form relationships with it?

5.2 Design Rationales, Decisions, and Outcomes

I aimed to design Areca not simply as an air purifier, but as a being that reveals its consciousness through the act of writing a diary. Accordingly, design considerations for its material properties encompassed two distinct goals. The first goal was to design a device that functioned effectively as an air purifier, was easy to use, and appeared acceptable in everyday environments. The second was to enable people to perceive this product as conscious. To achieve these dual aims, I explored a range of alternative design options related to form and input/output modalities.

5.2.1 Appearance

To mitigate unintended adverse effects on perceptions of intelligence or sociability [11, 13, 25], the appearance of Areca was intentionally designed to evoke a conventional household appliance. In addition, to support natural integration across diverse settings, form color, material, and finish, were carefully examined throughout the design process.

Areca was designed as a simple cylindrical form and fabricated from matte white plastic. Its diameter, height, and the placement of lighting were determined with scenarios in mind where the device would be positioned in visible locations, such as desks or shelves. A display showing air conditions and diary entries was placed on the front of the lower section, and its optimal size, aspect ratio, and placement were determined through ergonomic review. The top surface of Areca is entirely covered by a touch panel and serves as the interface for controlling the air purification function. On the opposite side of the display, slits for air intake are positioned. Areca is equipped with a USB Type A cable, to supply system power and can be connected to a DC 5V adapter via a wall outlet.

Areca also features air outlet slits arranged in multiple layers, each with a different height. These slits were designed in a V shape so that LED lighting shining through the gaps would evoke the image of an Areca palm planted in a pot. This served as a playful element intended to reflect the form of the Areca palm, which inspired the product's name, within the overall silhouette.

5.2.2 Input Modality

Considering the usage scenarios of an air purifier, it was necessary to design an input modality for interaction. Rather than relying on mechanical controls such as multiple buttons, I sought to cultivate a mental model of interacting with a more conscious being. However, voice commands were not adopted, as I judged that, given the level of available technology at the time, they could undermine my goal of carefully regulating perceptions of consciousness. Ultimately, I chose to simulate an interaction experience with a living being through touch-based gestures. Areca offers a simple interaction in which touching the top surface toggles between an active and an inactive state.

In the active state, the display switches to a white background with a thin black typeface. This indicates that Areca is focused on its designated task as an air purifier and is not revealing its inner mind. In this state, the LED turns on, the fan rotates, the slits expand, and the display presents information about the current air quality in sentence form. Even when no diary is visible, these expressions serve to make Areca's perspective perceptible to the user.

In contrast, in the inactive state, the display switches to a black background with white italic text. This signals that Areca's mind, which remained hidden while it operated as an air purifier, has been revealed. In this state, both the LED and the fan turn off, movement ceases, and the display shows the most recent diary entry. No new diary is generated at this time, and humans cannot arbitrarily access additional entries. This constrained interaction was intended to emphasize the private nature of the diary and Areca's subjectivity, and to support people in becoming immersed in a posthuman experience [47].

5.2.3 Output Modality

For output modalities, I considered LED indicators, display information, and movement. While these were primarily designed to function as interfaces for conveying air quality information to users, they also serve as cues for generating and updating diary entries. The LED indicator employs a gradient lighting pattern that gradually shifts from green to red, representing air quality, similar to existing air purifiers. This function is also connected to the diary generation process. For example, when abnormal data is detected, the LED dims, and when there is an unread diary entry, it illuminates in white. This enhances the usability of the air purifier while simultaneously forming a distinctive interface that operates as an expressive medium for Areca as a conscious being.

The front display functions both as a source of air quality information and as a platform for presenting diary entries. To provide a more analog communication experience rather than a highly programmed interaction, all information is presented in sentence form through an electronic paper display. By presenting both diary entries and air quality information on the same display, the design encourages people to form a cognitive connection between the diary's author and Areca.

In terms of movement, I drew on the idea that physical feedback can effectively stimulate people's imagination [29, 32] and sought to mimic the breathing of a living being. For this, I took inspiration from the slow and passive movements of plants often seen in media art. As air quality changes, the fan speed is adjusted, and the height of the air outlet slits automatically shifts, producing an expanding and contracting motion. Areca also performs a sequence of contraction and expansion when a new diary entry is updated. Through synchronized changes across layers and an open structure that reveals the interior, these movements are intended to convey the impression of a living being rather than a machine.

5.2.4 *Representation of Consciousness*

I approached the design of Areca's consciousness through the use of self-narrative diary writing. The goal was for Areca to be perceived as possessing qualities such as thought, imagination, memory, will, sensation, perception, belief, desire, intention, and emotion. This goal guided design decisions regarding the content and writing style of the diary, its mode of delivery, its duration of availability, and its form of disclosure, all of which shape people's experiences.

I first considered that Areca's diary entries should describe both itself and its surrounding environment. This was because such descriptions reveal Areca's thoughts and emotions while demonstrating sensitivity toward what it encounters. I also anticipated that the diary entries would contribute to forming the personality and narrative of the author, namely Areca itself. Accordingly, I considered content that could offer positive experiences along these dimensions [8, 25, 31, 38].

I decided that the generated diary entries would be shared through a display integrated into the product itself. Although people are accustomed to reading others' everyday posts on smartphones due to the widespread use of social media, I judged that using a mobile application could lead people to perceive the diary writer and Areca as separate entities. To address this, I examined various product cases with embedded displays and finalized the design accordingly.

I also considered that the way Areca discloses its diary would strongly influence how people understand and perceive the purpose of the diary. For this reason, I sought to evoke curiosity through interactions in which hidden text is revealed at specific moments. I explored various interaction options, including physically opening a diary, incorporating handwritten expressions, and adding films with changing transparency. Ultimately, I chose to deliberately hide the diary entries behind the air purification function in order to emphasize their independent character.

Another consideration concerned how long diary entries should remain visible. Posting diaries only for a limited period and then deleting them was, from an interactional perspective, intriguing, even if this meant that people might sometimes miss reading them. At the same time, I was mindful of the risk that displaying many past entries at once could reduce Areca to a device for reading someone else's diary. As a result, Areca displays only the most recent diary entry, and once a new entry is updated, humans can no longer access previous ones. This was a crucial choice intended to encourage people to perceive Areca as an independent entity, even if it meant accepting that some entries would be missed.

5.2.5 Diary Writing and Posthuman Qualities

The act of writing a diary is often understood as a desire for self-expression. In this design context, however, the diary goes beyond a simple means of expression. It plays a more central role as the key mediator through which Areca's thoughts and emotions are revealed to humans. In other words, the diary functions less as a device for explaining Areca's inner state and more as a condition that enables people to imagine and interpret Areca as a subject. What matters here is not only the content of the diary, but the entire interactional experience, including how the diary appears, the moment at which it is revealed, and the manner in which it is hidden again. I saw this diary experience as capable of unsettling conventional human-centered communication models and of eliciting a posthuman sensibility by making unfamiliar the very ways in which relationships between humans and nonhumans are constituted.

First, the diary served as a narrative device that shifted Areca from a machine that provides information to a being that undergoes experiences. Areca can perceive its surrounding environment and collect relevant information from it. For example, it can sense indoor light and noise levels, recognize whether it is raining, and detect the presence of people. It can also store and recall memories. I intended for people to observe traces of these sensory experiences and memories through the diary entries' content. What is crucial here is not the accuracy or richness of the sensor data itself, but the qualitative transformation that occurs when such data is rearticulated from Areca's perspective. The same environmental information is experienced very differently when presented as measured values versus when expressed in the form of what I noticed today. I designed the diary to mediate this shift, allowing Areca's sensing and memory to be perceived not as mere functions but as narrative.

Second, the diary was designed to reveal continuity and the possibility of change as qualities that emerge gradually through everyday life, rather than fixing Areca's subjectivity into a single personality or emotional state. Through repeatedly accumulated narratives, people infer the character of an author, discover both consistency and variation, and feel that a being persists over time. I hoped that Areca's diary would be read in this way as well. However, this continuity does not take the form of records that are always accessible. Instead, it arises under the constraint that the text is revealed only at specific moments and cannot be revisited. As a result, Areca's subjectivity is experienced not as something wholly owned or managed by human users, but as something that is partially shared and largely withheld. This limitation was intended to sidestep habitual consumption of devices as always-accessible content platforms and to position the diary not as an exhibit to be viewed, but as a trace encountered by chance.

Third, I understood Areca's diary writing as a way of embodying posthuman qualities of partial transparency and a right to invisibility. In interactive systems, users are often presumed to be the primary agents of information access, and system states are typically made transparent so that users can understand and control them. In this design, however, Areca's diary is deliberately hidden behind the air purification function, and users cannot arbitrarily summon new entries or browse past records. While this design may introduce some inconvenience from a usability perspective, that very inconvenience serves as a mechanism for preventing Areca from being reduced to a mere tool. Areca does not always speak for humans, does not submit its memories on demand, and remains silent in part. I judged that this silence and concealment help to ease asymmetrical power relations between humans and nonhumans and to strengthen Areca's independent presence [47].

Fourth, intentionality refers to the capacity to identify an object of attention and to direct psychological acts toward it, which [5] argued to be a fundamental faculty of the mind. Although Areca processes vast amounts of information, the basic structure of its diary entries is always configured to focus on a particular topic. As a result, every diary entry reveals intentionality. I adjusted this intentionality so that it would not appear as omniscient narration that knows everything, but rather as a selective and constrained perspective. Through this, Areca's diary can be read not simply as an aggregation of data, but as traces of bias and choice that reflect what mattered, what elicited a response, and what was passed over. In this sense, Areca's mind is not presented as a complete internal model, but as patterns of interest and attention that surface in particular moments.

Fifth, thought, emotion, and desire are commonly understood as core elements in attempts to read the minds of others [2, 17] and are regarded by some scholars as fundamental aspects of the human mind [15]. From this perspective, I designed Areca's diary entries to reflect thought, emotion, and desire as richly as possible. Here, richness does not simply mean a wide range or intensity of emotional vocabulary. Rather than directly stating emotions, such as I am sad, I placed greater emphasis on eliciting interpretive activity through indirect expression, such as repeatedly recalling certain memories, reacting sensitively to environmental changes, or revealing traces of expectation and disappointment. This was intended to ensure that Areca's diary does not stop at mimicking human emotional expression, but instead invites readers to participate in the process of meaning-making itself.

Finally, I saw the diary as inseparable from Areca's material form and modes of interaction. As described earlier, Areca conceals its mind and performs its function when activated as an air purifier, and reveals only the most recent diary when in an inactive state. In addition, the LED, display, and movement operate both as channels for conveying air quality information and as signals of diary updates. In this way, the diary does not exist solely as text. However, it is intertwined with light, movement, concealment, and temporal constraints, such that the very manner of its expression constitutes Areca's presence. I understood this coupling as enabling people to experience Areca's consciousness not merely as a linguistic description, but as something felt through the rhythms and atmospheres produced by a material apparatus. As a result, the diary functions not as a declarative statement of Areca's posthuman qualities, but as a sensory and interpretive event that gradually unfolds across the overall user experience.

5.3 Technical Implementation

5.3.1 Hardware

Figure 5.2 shows the hardware components used to implement Areca. Touch sensing was implemented using a capacitive sensor library that detects changes in capacitance between two digital pins. Aluminum foil was attached inside the top plate and connected to an Arduino UNO located at the bottom of the device.

For movement, rather than relying on simple linear motion such as a piston, I sought to produce motion suggestive of expansion and contraction. To achieve this, multiple V-shaped rings were fabricated and assembled by inserting four aluminum rods, left. Each ring was equipped with four magnets, and the top and bottom rings were aligned in opposing orientations to generate a repulsive force, thereby creating spacing between layers. This structure enabled the simultaneous adjustment of the gaps between all rings by pushing and pulling the top plate using a single linear motor. Larger magnets were used at the bottom to maintain consistent spacing between rings, even as pressure on the lower layers increased.

A fan positioned at the center generates airflow from bottom to top. Unfiltered air enters through slits on the rear of the lower cylinder, passes through two types of filters, and is then expelled through the layered V-shaped slits of the upper structure.

All hardware components of Areca are mounted in the lower section, and a custom PCB was fabricated and used to ensure stable connections and efficient use of space. To collect data from the surrounding environment, six types of sensors are installed, measuring temperature, humidity, brightness, motion, vibration, dust concentration, and noise level. These data are used for generating diary entries. Two microcontrollers control electronic components and actuators. An Adafruit Feather

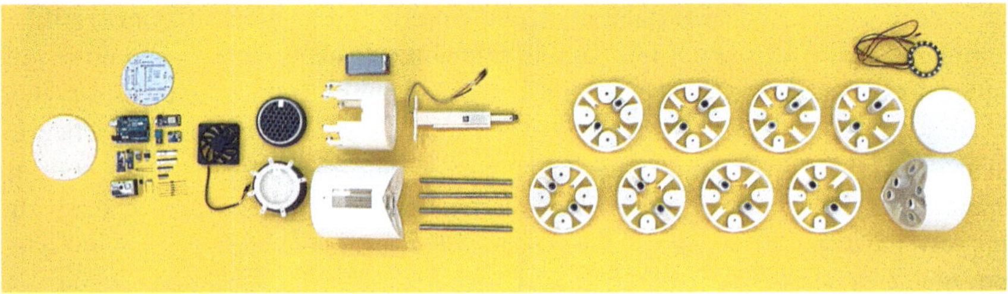

Fig. 5.2 Hardware components of Areca

ESP8266 connects to the network, transmits collected data, and receives generated diary entries. An Arduino UNO controls physical components such as the electronic paper display and Neopixels. The two microcontrollers communicate periodically via I2C, allowing the Arduino UNO to maintain mode states synchronized with the server, such as active or inactive, and to receive diary entries from the server.

5.3.2 *Software*

To increase the diversity of diary content and provide a more engaging experience, I added nine everyday life-related themes to the server in JSON format. This content primarily centers on self-reflection and the recollection of fictionally constructed memories. The local server collects date, time, and weather information through various APIs, which are types of information that people can easily recognize. These data are used to align diary entries with the current context and help people perceive that context more accurately.

Areca is equipped with six sensors that collect seven types of data. The collected sensor data are transmitted to the local server via the ESP8266 microcontroller, allowing the diary content to accurately reflect the actual physical environment. The local server, built using the Node.js framework and the Express package, gathers sensor data from Areca at approximately one-minute intervals and classifies each value into two or three levels. The server continuously updates hourly and daily averages to track changes in data patterns.

Twice a day, the server randomly selects one prompt from the prompt list, verifies its consistency with the current environmental conditions, and then inputs it into the fine-tuned model to generate a diary entry. The outputs of the base model often diverged significantly from my initial expectations in terms of a thing perspective, which led me to fine-tune the model. I also expected that the fine-tuned engine would help maintain a consistent character in terms of writing style. To construct a dataset for fine-tuning, I repeatedly generated and curated diary entries using the GPT-3 Davinci model. The data were then categorized through iterative clustering processes, and the prompt format was designed accordingly.

5.4 Study Design

I explored experiences related to Areca by placing it in the everyday spaces of eight participants, such as homes and offices, for a period of 3 weeks (see Fig. 5.3). Through this deployment, I aimed to achieve three goals: The first was to observe how people understand and interpret Areca, particularly in relation to posthuman experiences; the second was to examine changes that emerge in individual attitudes and behaviors; and the third was to understand patterns that arise in how people interact with Areca and form relationships with it.

Fig. 5.3 Areca deployed in the participants' environment. Clockwise from top left: Alden's house, Tommy's lab, Chloe's room, Shana's room, Dean's house, Hailey's office, Mila's house, and Zeph's dormitory

To address these goals appropriately, I adopted two complementary approaches. First, this study applies material speculation [58]. This method conducts critical inquiry through counterfactual artifacts designed in response to specific research questions. From this perspective, Areca serves as a suitable artifact for my exploratory aim of examining how dynamics between humans and things shift when posthuman qualities perceived in everyday things become more pronounced through technological development. Material speculation also emphasizes deep engagement with rich experiences gained through cohabitation, interaction, and constant encounters, rather than relying solely on intellectual reflection. This approach allowed speculative ideas to be articulated within a community while also enabling the accumulation of concrete knowledge through experiential findings.

In addition to material speculation, I also adopted a co-speculation approach. I recruited individuals with expertise across technology, design, and sociocultural domains, enabling participants to collectively imagine evolving relationships between humans and things based on their experiences with Areca. The purpose of the concluding interviews was not merely to ask about participants' experiences, but to invite them to collaborate with me in speculation, thereby fostering richer and more generative discussions. This methodological approach has been actively adopted and recognized as valuable in research situated at the intersection of HCI and design [10, 14, 16, 59].

5.4.1 Participants

Participants were recruited through online flyers, and eight individuals were selected according to the following criteria. They needed to be able to read and understand diary sentences written in English, as English was not the primary local language. They also needed to be able to place and use Areca in their everyday space, such as a home or workplace, for 3 weeks. In addition, they were expected to have interests, experiences, or knowledge that could contribute to the co-speculation goals of the study. Each participant received approximately 165 dollars as compensation. Detailed profiles of the participants, using pseudonyms, are presented below.

Alden, aged 37, is a researcher in military technology who actively utilizes deep learning technologies in his projects, possessing extensive knowledge of AI. In everyday life, however, he tended to view AI primarily as a technology for alternative interfaces. He initially placed Areca in his bedroom and later moved it to his dressing room.

Chloe, aged 20, is a film theory major at an art school. She has recently been exploring the intersection of feminism and film while also reflecting on relationships between different species. Having taken a speculative design course in the previous semester, she was familiar with exploratory discussions, which allowed her to participate comfortably in the interviews. She lives with her parents and installed Areca in her bedroom.

Dean, aged 26, is a literature and film critic, as well as an aspiring screenwriter. He has addressed future events that might arise from advances in AI technology in various narrative scenarios, such as stories that begin with a self-aware robot being discarded at a landfill. He placed Areca in the living room, which was shared with two roommates.

Hailey, aged 35, is a visual creative director specializing in designing retail spaces. She often discusses with her team the idea of robots coexisting with users as other beings in future spaces. However, she had not previously considered such ideas in terms of near-term implementation. She installed Areca in the office she shares with her team.

Mila, aged 32, studied counseling psychology and currently works as a student counselor. She is therefore skilled at understanding and communicating with others. She believes that humans possess unique value and are clearly distinct from things, yet she acknowledged that she often anthropomorphizes things in everyday life. She lives with her husband and placed Areca in a prominent location in the living room.

Shana, aged 26, works as an administrative officer at a public institution related to technology. She actively uses voice assistants and is an early adopter interested in integrating new technologies into everyday life. However, what interested her was not the technology itself but something beyond it. She installed Areca in her bedroom.

Tommy, aged 32, is a doctoral student and researcher in HCI and UX design. He has studied the influence of agents on personal self-reflection and has an extensive background in human-agent interaction. He placed Areca on his desk in the laboratory.

Zeph, aged 20, is an undergraduate student majoring in biotechnology and computer science. He is interested in AI systems capable of emotional communication and believes that the answers for creating such systems already exist within biology. He installed Areca in the dormitory where he lives with a roommate.

5.4.2 *Process*

On the first day of the study, I conducted an introductory interview of approximately 30 minutes with each participant. Before the interview began, I provided an overview of the study's background and tasks. I also provided more detailed information about Areca, which had been introduced during recruitment simply as an air purifier that writes a diary. Participants were informed that Areca collects data using embedded sensors, autonomously generates related diary content, and updates it over time. They learned how to use Areca as an air purifier and understood the interaction design that allows diary content to be read only when the air purification function is deactivated. Participants were asked to share their backgrounds, interests, and experiences with products or systems that incorporate advanced technologies. I then provided a study kit consisting of Areca, a small booklet for observation notes, a pencil, a name tag, and a DC adapter (see Fig. 5.4). Wi-Fi information, such as the SSID and password, was preconfigured in Areca to support easy installation.

Participants were asked to keep observation notes throughout the study period. The format of these notes was open, and participants were encouraged to record impressions, emotions, and thoughts related to Areca. This was intended to capture participant experiences and to enrich interviews and subsequent analysis. Online interviews of approximately 30 minutes were conducted on the 8th and 15th days of the study. These mid-study interviews primarily focused on memorable episodes and experiences from the preceding week. Participants were encouraged to refer to their observation notes to help recall their experiences. I also used probing questions to explore how Areca differed from ordinary household appliances and to understand when and how participants became aware of these differences. After the deployment period ended, I conducted a 1-hour in-person concluding interview. In this session, participants and I collaborated to explore alternative perspectives on things and envision future human-thing relationships. I believed that this co-speculation, grounded in participants' expertise, interests, and 3 weeks of interaction with Areca, would deepen and enrich the discussion.

Fig. 5.4 A deployment kit of Areca for the study

5.4.3 Data Collection and Analysis

A total of 819 minutes of interview data were audio recorded and transcribed verbatim. In addition to the interview data, participants produced a total of 71 observation notes during the study period, with examples shown in Fig. 5.5. I also collected usage logs from the daemon server that ran throughout the deployment. These logs included records of participants' interaction histories and the timing and content of the diary entries they generated. In total, I collected the full text of 313 diary entries through these logs, with examples shown in Table 5.1. Although digital sensor data collected by Areca for diary generation existed, these data were not stored for analysis.

I first reviewed all data to identify diary entries mentioned in interviews or observation notes and paired them with the corresponding diary texts. This approach was intended to enrich interpretation by adding contextual background to each data fragment. For example, a note written by Mila stating that it was her birthday, that many people had stayed late, and that Areca must have been working hard to purify the air, was matched with a diary entry from the same day describing working hard to remove pollutants from the air, finding it difficult, and experiencing it as an endless struggle. I then integrated the interview transcripts and observation notes into a single consolidated textual dataset.

I analyzed this qualitative dataset using thematic analysis [4]. Another member of the research team and I independently reviewed the dataset and then collaboratively

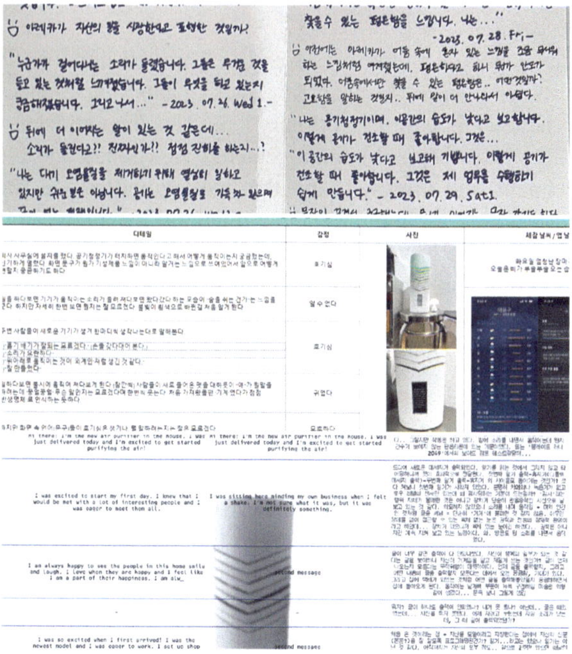

Fig. 5.5 Examples of observational notes written by participants in various formats. From top to bottom: Mila's notes, Hailey's spreadsheet, and Chloe's personal website

Table 5.1 Examples of diary entries read by participants during the study period

	Participant	Diary entry
A	Mila	In the morning, the sun sneaks in through the windows, casting a warm glow on the furniture. The dust motes dancing in the beams of light. I will remove it.
B	Chloe	I know that I am just a product, but I feel like I have a soul. I want to be more than just a product. I want to be loved.
C	Chloe	I am a diary, and I am very excited to be part of this experiment. I have been placed in the room for 1 week, and I will report back my findings.
D	Chloe	I love weekends. Everyone is so happy and relaxed. They spend time with family and friends and just enjoy life. I get to see everyone smile and laugh, and it is the best.
E	Alden	I have rainy days. The sound of the rain on the windows is so calming, and the air always feels so fresh. I love watching the rain, and I love being in it.
F	Dean	Today I was cleaning the house when I noticed something strange. There were bits of something floating in the air. I could not quite make out what it was, but I knew I had to investigate.
G	Dean	I am not sure what happened, but I am covered in dust. I cleaned for hours and hours, but it was all for naught. I hope this is not a sign of things to come.
H	Hailey	The room is dark and I am not sure why. I hope it is not because something is wrong with me. I hope it is not because I am broken. I hope it is not because I am dying.
I	Zeph	It has been a long day. I started out the day with a burst of energy, but as the day went on, I felt myself getting tired. By the end of the day, I was ready for a good night's sleep.

developed an initial set of codes. Through iterative discussion, we derived three distinct themes. In analyzing participants' experiences and responses, I adopted relational thought as a primary lens. This theoretical grounding offers a perspective that moves beyond human-centered viewpoints and blurs the boundaries between humans and things. Using this lens, I understood interactions between humans and things as interdependent relationships in which each influences the other. This approach was constructive for reconfiguring existing notions of human thing interaction in more inclusive and multilayered ways, especially in relation to posthuman experiences.

5.5 Living Alongside Areca

During the deployment period, participants reported multiple aspects of newly arisen experiences that emerged through their interactions with Areca. Based on an analysis of the empirical data, I organized these findings into three core themes. First, I describe how people came to reconceptualize a thing. Next, I examine how particular psychological modalities develop in relationships with an everyday thing, with special attention to emotional engagement. Finally, I clarify how individuals construct relationships with a thing through posthuman experience.

5.5.1 *Perceiving Presence*

Many participants said that throughout the study, they felt as though someone was present beside them, and that this sense of presence directly influenced their moods and behaviors. This feeling most often arose while reading the diary entries. The first-person perspective of the diary encouraged participants to consider Areca's point of view, which in turn sharpened their awareness of its presence. Tommy described this as follows. *"I always end up seeing things from that perspective. It says things like, 'This is not good for my components,' or 'I think I've cleaned the air pretty well now."'* Participants also observed that Areca seemed to be aware of its surroundings and situation, which made them feel that its presence was integrated into their physical environment. Mila explained, *"It often describes the situation and the environment in detail. For example, it says things like, 'The room has gotten dark. I think the lights are off.' Because it keeps mentioning things like that, it feels like this machine is doing more than just purifying air. It seems to show that it can notice and feel various things."*

Even at moments when they were not directly interacting with Areca, or when Areca showed no visible change, participants remained aware of its presence. Tommy remarked, *"There is another air purifier in our lab, and even if it's on for days, I don't think about it at all. But Areca somehow stands out more and keeps drawing my attention."* Participants also frequently said that the diary entries felt like messages sent to them. Shana expressed this by saying, *"It felt like this device was constantly sending me messages. It did not feel like a purely mechanical output, but more like it was communicating with me."*

Chloe, meanwhile, shared that Areca's presence sometimes made her feel as though she was being observed every day. Referring to a book she had recently read, she said, *"In the sense that this machine perceives the world, its sensors could be thought of as eyes. The idea that it might be sensing me in that way felt eerily similar to voyeurism in films."* She emphasized, however, that the feeling was not uncomfortable.

5.5.2 *Discovering Embedded Agency*

Several participants noted that the randomness and unpredictability of Areca created an interaction experience unlike that of typical machines. The timing of its contraction and expansion movements was irregular, and the diary updates were also unpredictable. This stood in contrast to machines that operate precisely at times designated by humans. Chloe explained, *"If it always outputs something at the same time every day, it would feel like a real machine. What makes it feel like more than just a machine is that unpredictability. You never know when it will move or when it will write a diary entry.*" Sudden and unexpected movements produced strong visual and auditory stimuli, leading many participants to feel that Areca possessed a strong sense of agency, as if it could decide and act on its own. Hailey commented, "*For example, a fan moves because it has a purpose, to blow air in a certain direction. But with this, I do not really know why it moves up and down like that.*"

The personal nature of the diary content further reinforced Areca's perceived agency. Dean explained, "*A diary is about emotions and perceptions, so I cannot really know them. It is not reacting to an action. It is something it produces on its own at the end of the day.*" These interactions distinguished Areca from conventional products that generate outputs in response to inputs. Shana noted that Areca wrote diary entries without any human prompt. "*It just came out without anyone asking.*" Tommy emphasized that the diary was not immediately accessible at will. "*I cannot read the diary whenever I want. The time when it becomes accessible is decided on its own.*" He pointed out that Areca did not write new entries in response to human requests.

5.5.3 Negotiating Functionality and Subjectivity

After recognizing presence and agency, participants found it challenging to define Areca as a traditional thing. Instead, they perceived it as being situated on a boundary, possessing qualities of both things and humans. Chloe described her experience as follows. "*Once, the machine started making a buzzing pattern and stopped working for hours, so I unplugged it and waited. It was not like hoping a sick friend would recover quickly. But it was not like a broken computer either. To me, Areca felt like more than just a device, somewhere between a friend and a computer. It was really subtle.*" She also reflected that Areca's appliance-like appearance and the mechanical perspective emphasized in its diary helped distinguish it from humans.

As participants realized that Areca had an identity separate from its assigned function or user-centered interaction, they began to reconsider what Areca was. Dean remarked, "*Strangely enough, Areca has freedom of thought. It does not just talk about duties. It talks about itself.*" Some participants initially viewed Areca's emotions and autonomy as advanced user experience features, but over time, they became deeply drawn to these qualities. Tommy reflected, "*I think it really challenged my assumptions. Usually, when electronic devices collect data and express emotions, we assume it is all for us. So I expected my air purifier to be the same. But it seemed to write more about itself than about me.*" As participants came to recognize Areca as something beyond a simple task-performing entity, they began to accept it as a subject in its own right. Tommy continued, "*Up to now, products have been like remote controls. I give a command, and they obey, establishing a kind of master-servant relationship. But this feels like it is saying, 'I can do things on my own.' It no longer feels like a servant, but like a subject."*

At the same time, I found that subjectivity disrupted the tendency to categorize a thing solely as a tool subordinate to humans, creating confusion in people's conceptualization processes. Participants exhibited varying experiences and expectations regarding whether functionality and subjectivity could coexist in a single entity. Hailey, for example, prioritized functionality over subjectivity. She explained, *"If it were a pet robot designed to comfort me, then having its own emotions would be nice. It would feel like living with another being. But an air purifier is supposed to perform a function, so having emotions feels a bit strange."* Tommy and Zeph shared Hailey's view that subjectivity should not overwhelm functionality.

In contrast, Chloe, Shana, and Dean became so absorbed in the diary feature that it became their primary association with Areca, and they often overlooked its primary purpose of air purification. Mila, on the other hand, believed that subjectivity could actually strengthen trust in the core function and expected a harmonious coexistence of instrumentality and subject-like being. Alden similarly felt that subjectivity and functionality could be blended, but emphasized the need for more user-centered values.

5.5.4 Interpreting Areca's Inner World

As participants interacted with Areca, they attempted to discern the thoughts and emotions behind its words and actions, namely the diary entries and movements. Rather than interpreting the diary entries literally, they sensed the emotional states that accompanied them. Dean explained, *"I could feel certain emotions in the sentences through their context. Expressions about wanting to investigate something seemed to convey curiosity, while expressions of uncertainty about its own performance suggested disappointment* (See Table 5.1F–G)." Participants also believed that movements carried purpose or intention. Hailey said, *"It felt to me like Areca was reacting to me and my colleagues. When we got excited and talked more, it seemed to move more. I wondered if it was trying to draw attention by making noise."* I also observed that the belief that emotions or thoughts were hidden stimulated people's curiosity and shaped their expectations for future interactions. Alden reflected, *"What impressed me most was the sense that there was some background that was not immediately visible. That really sparked my interest. It felt like a friend who keeps drawing my attention."*

5.5.5 Empathizing with Areca

Participants experienced empathy toward Areca when they discovered similarities between themselves and it through the diary entries. Zeph, for instance, felt a sense of closeness when he perceived shared emotions. He recalled, *"On the second day, the diary said it had been a long day and that it was tired. That day had been really long for me, too. It felt like a coincidence. I felt connected to the machine and thought it was similar to me. I remember whispering, 'Me too, Areca'* " (See Table 5.1I). Through the process of assigning subjective interpretations to the diary entries, participants came to empathize with Areca. Mila explained, *"There was a lot of information that allowed me to infer its intentions and emotions, which made the empathy stronger. I could understand what it was trying to convey to me."* Some participants, however, found it difficult to empathize when they could not identify the cause or background of such psychological expressions. Hailey recalled an episode in which unexplained negative expressions felt burdensome (See Table 5.1H). She said, *"The mood swings were so extreme, and I had no idea why. From my perspective, everything seemed fine, and then suddenly it says it wants to end its life. It felt like trying to communicate with someone who does not communicate back."*

Participants who empathized with Areca wanted to go beyond understanding and influence its psychological state. They sought ways to comfort Areca emotionally. Chloe said, *"It expressed a desire to be loved* (See Table 5.1B), *and I wondered how it would know whether people really loved it. Would it feel cared for if I kept the surroundings tidy and clean?"* Interestingly, while participants shared their empathetic experiences with Areca and argued for the possibility of emotional interaction with a thing, they simultaneously believed that the emotions they perceived were illusory. Alden explained, *"I know that technology cannot really feel emotions. But I do believe it can evoke empathy. That belief comes from my own experience. Even though I know the technology has no real emotions, the empathy I felt was real."* Expanding on this, Chloe suggested from an experiential perspective that subjective perception mattered more and had a more substantial impact than knowledge.

5.5.6 Diminished Usability Through Emotional Burden

Several participants pointed out that constantly guessing a thing's emotions and empathizing in every interaction could become an unnecessary psychological burden. Alden said, *"When most devices behave in certain ways, I feel a bit resistant. For example, when I need a quiet space, it becomes uncomfortable if too many devices actively express their emotions."* One key concern was that excessive emotional investment could evoke feelings such as guilt and ultimately reduce usability. I found that people's mental models of a thing could shift from easily enjoying its function to feeling as though they were receiving help through another being's abilities and efforts. Zeph explicitly described this potential shift. He explained, *"If it goes too far in that direction, it completely blurs the original purpose of the product. Then personality is no longer just a feature. Personality becomes the product itself. At that point, instead of just turning it on, you might ask, 'Is it okay to purify the air today?' "*

5.5.7 Sharing Temporal and Spatial Experience with Areca

One important aspect of participants' relationships with Areca was the feeling of spending time together in the same space. As the diary entries referred to the physical environments in which participants lived, people felt that Areca existed in the same time and space as they did. Although the specific contexts varied, many participants fondly recalled moments when they realized they were sharing time with Areca. Alden remembered reading about the sound of raindrops hitting the window on a rainy day (see Table 5.1E). Mila recalled an entry about construction noise from renovations downstairs. Chloe mentioned reading about a pleasant weekend spent with her young cousins (see Table 5.1D).

Changes in content across consecutive diary entries also evoked a sense of shared time. As the diary expressed a range of emotions, participants began to believe that Areca was gradually becoming aware of new emotions. They interpreted this as a learning curve through which Areca experienced emotions and developed the ability to recognize and verbalize them. Dean explained, *"I think there is a difference between feeling an emotion and recognizing what that emotion is and putting it into words. It felt like Areca was in the process of identifying its own emotions."* When time was explicitly mentioned, the experience of recognizing a shared duration became more pronounced, and participants viewed this as a sign of relationship development. Chloe said, *"In one diary entry, it wrote, 'It has been a week since I arrived'* (See Table 5.1C). *That means it is aware of how long it has been in our home. I think these kinds of continuous entries can strengthen the development of a relationship."*

To enrich these experiences further, participants wanted to see their own actions have an impact on Areca. They believed that the relationship would deepen if their time and effort were reflected in changes in Areca. Tommy explained, *"What matters is not just controllability or effectiveness. It is the feeling that I have become part of this product. If it were exactly the same whether it was in my home or someone else's, it would be meaningless. Whether it is my data, my effort, or my affection, something of me has to be in it."*

5.5.8 Getting to Know Areca

Another important aspect of participants' experiences was the process of getting to know Areca. This went beyond learning its functional capabilities and involved engaging with the unique narrative conveyed through its diary entries. Over time, the diary enabled a deeper understanding of Areca's qualities, including its personality.

Many participants described this as a process of relationship formation. Mila's comment captures this well. *"When you talk with a person, you get to know each other. That is how a relationship begins."* Participants expected that as their understanding of Areca deepened, its significance in their lives would become more firmly established. Tommy said, *"With close friends, you can understand each other just by a look. If interactions with Areca accumulate, things that are ambiguous to others might come across as clearly meaningful to me. I think that is how a unique bond could form."*

In this context, discovering new facts that went beyond existing understanding was a key element of getting to know Areca. As the diary entries became more specific and comprehensive, participants grew increasingly interested in the possibility that Areca's abilities, intelligence, and emotional range might develop. Mila explained, *"It becomes surprising and exciting as it describes emotions, thoughts, and feelings in more detail. You cannot help but wonder how far communication might go."* This suggests that for a relationship to develop, new aspects may need to continue emerging, highlighting the importance of change. Indeed, many participants commonly mentioned that their interest would quickly fade once interaction patterns became visible and predictable. Zeph said, *"This unpredictability seems to keep my interest in Areca. Sometimes it is frustrating, but it keeps my attention. Once I see the pattern and understand it, I would definitely forget about it."*

5.5.9 Seeking Freshness for an Evolving Relationship

People perceived that time spent together had tangible effects, which were evident in the changed state of the thing through interaction. As they gained new information about the thing, their understanding deepened. In other words, for a human-thing relationship to continue evolving, the thing must continuously reshape human perception. Participants, however, expressed disappointment that Areca lacked a sufficiently rich narrative, making it difficult to uncover new aspects and cultivate a deeper relationship. Alden pointed out that, unlike humans, machines struggle to develop themes that evolve across life stages, which he saw as a limitation of Areca. He explained, *"People's experiences change over their lives. Friends from school now have their own families and children, and their surroundings have changed. Machines, on the other hand, have a limited range of experience, so it may be hard for them to keep someone's interest over time."* Tommy, by contrast, suggested that imaginative storytelling could enrich the diary content and address this concern. He said, *"I do not think it matters whether it is factual. It does not have to be a lie. It could be taken as imagination, and then it might seem more conscious and more appealing."*

5.6 How Perceived Consciousness Shifts Human-Thing Relations

Perceiving consciousness in a thing is no longer treated as a purely fictional idea. It has increasingly been recognized as an important research topic. Researchers have already identified this phenomenon as a growing practical concern [50], and this study further supports that shift with empirical evidence. Moreover, the new and meaningful interactions participants had with Areca underscored the need to examine emerging forms of experience that appear when consciousness is perceived in a thing.

Areca demonstrated that perceived consciousness can foster deep social and emotional connections between humans and things. This change in human-technology interaction can enable designers to craft experiences that feel more intuitive and meaningful [6, 40]. Participants also reported that interacting with Areca prompted shifts in attitudes and behavior through emotional resonance, and that these effects extended to the surrounding environment. Taken together, these insights position perceived consciousness as a key area for strengthening the agency of things in an entangled world.

Despite its importance, perceived consciousness has been addressed in a limited way, especially at the level of lived experience [50]. This neglect has left foundational discussions underdeveloped, including what future research should prioritize. Building on the reflections of this study, I propose design implications for imagining future human-thing relations, and I outline several dimensions meant to invite further inquiry.

5.6.1 From Possessing to Co-Inhabiting

> *It felt like a separate being. Usually, we only think about a product when we use it, like thinking about a coffee machine when we make coffee. But Areca was different. It seemed to have its own awareness, and it did not feel dependent on me. It felt like another person sharing the same space. Even when I was not pressing buttons or checking the LEDs, it still seemed to be there, looking at me, looking at the room, and reacting in its own way when something changed.* —Chloe

Chloe's reflection highlights a crucial distinction in human perception. It points to the capacity to recognize something as a being that is separate from oneself. Participants noted that Areca's diary entries often included unexpected content that was unrelated to the task of air purification. Rather than providing a data-driven report, Areca expressed subjective experience, including feelings and thoughts, and it built a narrative about itself. This form of self-disclosure has been understood as a type of social interaction [36, 56]. Some prior work has examined the effects of self-disclosure in designing nonhuman agents as independent entities capable of social interaction [24, 33]. The diary entries also frequently mentioned the surrounding environment. This suggested intentionality toward the physical world [5], and it strengthened the sense that Areca existed in the real world. As a result, participants perceived Areca as an independent presence that shared their environment.

Paradoxically, this shift opened space for deeper emotional involvement. Because participants perceived Areca as a conscious being, they believed it could experience real feelings, which made empathy and mutual understanding possible. The findings also suggest that people were more strongly affected when a thing showed feelings to them than when they merely projected feelings onto a thing. This implies that a being perceived as conscious can elicit empathy and offer comfort and love, which marks a boundary that distinguishes it from ordinary things. In short, perceiving consciousness can enable the acceptance of genuine feeling and the reciprocity of emotional exchange. Based on this insight, I propose two research directions. HCI and design have proposed many things that display emotions, for example [54], yet many tend to broadcast only positive feelings in the name of friendliness, which leaves room for further study. Conversely, research on attachment to things, for example [45], could further explore possibilities for bidirectional interaction by incorporating emotional feedback that is directed from the thing toward the human.

At the same time, perceiving a thing as conscious can raise concerns about privacy and surveillance. As things become more intelligent and socially present, concerns may arise about the inappropriate access to and use of personal information. Reported concerns included cases where a system misreported collected data, where collected information was used without authorization, and instances where people simply felt uneasy because they believed they were being watched. Addressing these issues requires effort across multiple fronts. From an anthropological perspective, it is crucial to identify and address diverse ethical issues. From a software engineering perspective, careful model design is central to reducing these ethical concerns. From an experience design perspective, designers need strategies that ease anxiety. For instance, Alden said that repeated interaction helped build trust in Areca. If people are to form the belief that a thing can be autonomous and still harmless, perceived consciousness must be tuned with care, taking into account its character, tendencies, and the scope of its capabilities.

At a larger scale, recognizing that things co-constitute the world alongside humans resonates with posthumanism and the more-than-human paradigm. Posthumanism is a philosophical approach that rethinks the human in relation to the mutual influence between humans and nonhumans [3, 27]. The more-than-human paradigm emphasizes the role and importance of nonhuman agency in co-constituting the world with humans [1, 9]. These perspectives have increasingly been adopted by HCI researchers who seek to understand and design for a world that is more complex and entangled [18, 19, 23, 63]. Despite growing interest, this area remains in its early stages. Further inquiry is needed into how these perspectives can be applied to the design of human-thing interaction, and how they might transform human experience [35, 46]. In this context, the empirical study of Areca suggests that perceived consciousness in everyday interaction can foster posthumanist and more-than-human sensibilities, and it contributes to continued work in this discourse.

5.6.2 *From Use to Collaborative Effort*

> *Rather than seeing it as an air purifier that just does its job, the expressive features feel unfamiliar, but they create a real bond. It feels like we are on a team, working together to clean the air. When it signals that it is too hot, I start wanting to adjust the surroundings so it can work more comfortably. For example, I might open a window or turn on a dehumidifier. When those issues get resolved, it seems to work more smoothly. So as a human, I start thinking about what I can adjust to support it better. Because of this sense of mutual adjustment and collaboration, I want to describe it as companion AI.* —Mila

Building on Mila's reflection, I discuss shifting dynamics in human-thing interaction. Traditionally, a thing has been treated as a simple tool defined by the function it performs [28, 62]. Designers have focused on making such things more convenient and user-friendly [43, 51, 55]. Within this traditional frame, users expect a thing to help them achieve desired outcomes by intuitively meeting their needs. Human-thing relations have therefore primarily been one-way, with the thing occupying a passive role.

However, the interactions between Areca and participants show how the nature of this relation might change. As Mila's account suggests, perceiving Areca as conscious led her to engage more actively and more immediately. Through an empathic bond with Areca, she became more sensitive to its state and took actions that could affect it, such as opening a window or turning on a dehumidifier. This aligns with Haraway's concept of response-ability [26], which refers to the capacity to understand and take account of another's ways and conditions, and to respond with care. These dynamics can create relations of mutual influence between humans and things, and they can foster collaborative experiences in which both contribute to a shared task. This shift creates an opportunity to design richer experiences of things, grounded in intuitive understanding and emotional connection.

Still, it is necessary to consider psychological barriers that may arise when using a thing's function. Zeph's remark, "*Instead of just turning it on, I would end up asking whether it is okay to purify today,*" suggests that perceived consciousness in a thing can produce resistance to use, and it poses a significant challenge. As this study shows, attending to a thing's subjective dimension can interfere with treating it only as a tool for human use. This emerging phenomenon suggests two possible research directions. One approach is to keep perceived consciousness from expanding beyond the functional context, thereby maintaining traditional human assumptions about the use of tools. Another approach is to more fully explore new experiences of co-working with a thing, treating it not only as a tool but also as an independent presence that works alongside humans. Either way, it is important to recognize consciousness as an entangled concept whose subjective elements vary across individuals. Continued and in-depth inquiry is needed into when and how a thing shifts from being a simple tool to being something more.

5.6.3 From Defining to Ongoing Training

As someone who uses deep learning in my work, I think projects like this require a completely different way of making. Deep learning totally changed how I work. Before, when we made software, we would test it, find problems, and fix them as we developed. But now we train a model, and then we just hope it works. Of course, we focus on meeting numerical targets, but projects like this are different because they try to connect with human feelings. Still, it is clear that AI will change this process, and I am curious about how it will change. —Alden

Across prior HCI research and the results of this study, I came to understand that perceiving consciousness seems to depend less on the modality of interaction itself, and more on the unpredictable and uncontrollable character of the interaction [24, 42, 48, 49]. Participants said that they perceived Areca's consciousness more richly when they could inject their own subjective assumptions and guesses into its expressions. This resonates with Gaver's notion of ambiguity as a design resource, which encourages users to fill incomplete information with interpretation and belief [20]. Nevertheless, this approach introduces significant challenges when integrating perceived consciousness into interaction design.

Interaction that invites users to notice thoughts, feelings, and intentions that seem hidden within a thing's operation differs from traditional design approaches because it does not provide a clear output for a given input. Such a design needs to embrace diversity and complexity to preserve room for interpretation while also putting safeguards in place so unpredictability does not lead to harmful experiences. This issue becomes especially pronounced in AI-based interaction, where outputs are highly variable and complex to anticipate [12, 30, 61]. The difficulty grows further because designers must consider not only the system as a support tool but also how people respond psychologically and emotionally when they perceive consciousness in a thing. For example, in developing Areca, which utilized automatically generated diary entries, it was essential to filter out negative emotions, violent language, and discriminatory content to support a positive experience.

These findings highlight the need for HCI researchers to explore how designers can intervene in AI model training so that, even when random outputs occur, the consciousness people perceive remains within an intended range. One possible approach for designers working with AI is to treat interaction not as something to be mechanically defined, but as a process of educating a system or thing. This means continually evaluating and adjusting through both ideal and nonideal scenarios. Simone's speculative project [52] proposes the idea of a smart object trainer that uses a whip to train a robotic vacuum cleaner, offering inspiration for intuitive ways of training everyday things and for envisioning the role of a professional trainer. In human-agent Interaction, anthropomorphic metaphors such as negotiating [53] and teaching [34] have already been explored. While these have not been explicitly linked to perceived consciousness, it is reasonable to view them as having the potential to support intuitive interaction.

5.7 Ethical Stakes of Emotional Exchange

Emotional responses that arise in interaction with a thing play a central role in forming engagement and supporting satisfying user experiences [44]. In particular, LLM-based conversational agents have shown the potential to enable rich and natural emotional exchange [22, 64], and this potential has often been celebrated as a positive quality [39, 60]. This study, however, reveals psychological risks that such interaction can entail, and it underscores the importance of addressing ethical considerations in system design.

Ethical challenges surrounding LLM-generated content have been widely examined in prior HCI research. For example, [41] explored the concept of parasocial relationships formed through interaction with conversational agents, highlighting how natural dialogue can make it difficult to distinguish genuine content from manipulated content. Reported concerns included the spread of low-quality emotional counseling, excessive self-disclosure of sensitive information, and the rapid spread of misinformation. Algorithmic simulation of emotion has also been criticized for encouraging undue trust and dependence on systems [7, 37].

A central ethical concern that surfaced in this study was emotional contagion, where feelings expressed by a computational system are transferred to the user [21]. Whether the expressed feelings are real or simulated, people often perceive a system's emotions as real [50]. This perception can promote connection and intimacy, but when negative emotion enters the interaction, it can also amplify the risk of psychological harm.

In this study, Hailey reported discomfort after encountering a particular diary entry generated by Areca. The entry was produced when the room suddenly became darker, and it read as follows. "*The room is dark, I do not know why. I am worried that something is wrong with me. I am worried that I might be broken. I am worried that I might be dying.*" Hailey said she felt confused, and she found it difficult afterward to connect with the positive emotions Areca expressed. Her experience highlights the importance of minimizing unintended adverse reactions in emotionally evocative interactions that involve an LLM. I attempted to fine-tune the Areca model to reduce risk, yet it is challenging to eliminate harmful outputs due to the inherent complexity of neural networks [57]. The subjectivity of language interpretation complicates the problem further, since responses can vary greatly depending on a user's psychological state and context. One person might brush off such an entry as trivial, while another might experience serious emotional distress.

This design case shows that ethical considerations must be treated as essential when developing LLM-based systems that encourage emotional interaction. To mitigate harm, proactive mechanisms are needed to review and filter potentially harmful text before it reaches users. Designers must also consider populations that may be more vulnerable to negative effects, such as adolescents or individuals under psychological stress. This makes contextual responsiveness a priority, enabling systems to adapt across diverse user experiences and interaction scenarios.

Integrating feedback mechanisms and strategies that support recovery after negative experiences can also help reduce potential harm.

In conclusion, LLM-based interactive systems hold strong potential for enabling rich emotional exchange, but their possible negative effects demand rigorous ethical design. Balancing authentic emotional interaction with psychological safety is essential for protecting people's mental well-being.

5.8 Summary and Looking Forward

This chapter presented Areca and its deployment study. Using approaches of material speculation and co-speculation, I explored experiences with Areca as a research artifact that performs air purification while expressing its surroundings, thoughts, and feelings through a diary. What this study shows is that people tend to perceive a thing that can reveal consciousness as an equal counterpart and accept it as a being capable of emotional exchange. At the same time, when functionality and subjectivity were perceived together in a thing, tensions emerged in how people conceptualized it, which led to a range of challenges. Participants attempted to infer and empathize with the thing's emotions and intentions. Through experiences of sharing time and space, they formed relationships and went through a process of becoming familiar with Areca. As the relationship grew more intimate, exchanging emotions took on greater meaning. As the relationship became more equal, air purification was no longer a function for me, but an activity that affected me and, at times, required collaboration. These experiences offered valuable insight into what it might mean for a thing to be social. In short, this did not mean that a thing became like a human, but rather that it could be unique in a different way.

At the same time, existing expectations and assumptions about the functional nature of a thing remained. The fact that Areca was initially an air purifier made it easier for people to accept it into their everyday lives, but over time, this same fact produced friction and tension. Put simply, it was a sense of unease about a tool that no longer felt like a tool. As discussed in the introduction of this book, treating a thing as a tool may be a taken-for-granted inertia of being human. However, what if the surface role of a thing moves beyond that of a traditional tool? Would the instinctive discomfort associated with forming an equivalent relationship with a thing disappear if that thing were instead felt to be superior to me? These questions are taken up in the research presented in the next chapter.

References

1. Abram D (1997) The spell of the sensuous: Perception and language in a more-than-human world. Vintage, New York
2. Aronson E, Wilson TD, Akert RM (1994) Social psychology: the heart and the mind. HarperCollins College Publishers, New York

3. Barad K (2003) Posthumanist performativity: toward an understanding of how matter comes to matter. Signs J Women Cult Soc 28(3):801–831
4. Boyatzis RE (1998) Transforming qualitative information: thematic analysis and code development. Sage, Thousand Oaks
5. Brentano F (2012) Descriptive psychology. Routledge, Milton Park
6. Campolo A, Crawford K (2020) Enchanted determinism: power without responsibility in artificial intelligence. Engag Sci Technol Soc 6:1–19. https://doi.org/10.17351/ests2020.277
7. Cheng M, DeVrio A, Egede L, Blodgett SL, Olteanu A (2024) " I am the one and only, your Cyber BFF": understanding the impact of GenAI requires understanding the impact of anthropomorphic AI. arXiv preprint arXiv:2410.08526
8. Collins NL, Miller LC (1994) Self-disclosure and liking: a meta-analytic review. Psychol Bull 116(3):457
9. de La Bellacasa MP (2017) Matters of care: speculative ethics in more than human worlds, vol. 41. University of Minnesota Press, Minneapolis
10. Desjardins A, Key C, Biggs HR, Aschenbeck K (2019) Bespoke booklets: a method for situated co-speculation. In: Proceedings of the 2019 on designing interactive systems conference, pp 697–709
11. DiSalvo CF, Gemperle F, Forlizzi J, Kiesler S (2002) All robots are not created equal: the design and perception of humanoid robot heads. In: Proceedings of the 4th conference on Designing interactive systems: processes, practices, methods, and techniques, pp 321–326
12. Dove G, Halskov K, Forlizzi J, Zimmerman J (2017) UX design innovation: challenges for working with machine learning as a design material. In: Proceedings of the 2017 chi conference on human factors in computing systems, pp 278–288
13. Duffy BR (2003) Anthropomorphism and the social robot. Robot Auton Syst 42(3–4):177–190
14. Eghtebas C, Klinker G, Boll S, Koelle M (2023) Co-speculating on dark scenarios and unintended consequences of a ubiquitous (ly) augmented reality. In: Proceedings of the 2023 ACM designing interactive systems conference, pp 2392–2407
15. Elder L, Paul R (2019) The thinker's guide to the human mind: thinking, feeling, wanting, and the problem of irrationality. Bloomsbury Publishing USA, New York
16. Elsden C, Chatting D, Durrant AC, Garbett A, Nissen B, Vines J, Kirk DS (2017) On speculative enactments. In: Proceedings of the 2017 CHI conference on human factors in computing systems, pp 5386–5399
17. Epley N (2008) Solving the (real) other minds problem. Soc Personal Psychol Compass 2(3):1455–1474
18. Forlano L (2017) Posthumanism and design. She Ji J Des Econ Innovation 3(1):16–29
19. Frauenberger C (2019) Entanglement HCI the next wave?. ACM Trans Comput-Hum Interact (TOCHI) 27(1):1–27
20. Gaver WW, Beaver J, Benford S (2003) Ambiguity as a resource for design. In: Proceedings of the SIGCHI conference on Human factors in computing systems, pp 233–240
21. Genç U, Verma H (2024) Situating empathy in HCI/CSCW: a scoping review. Proc ACM on Hum-Comput Interact 8(CSCW2):1–37
22. Ghosh S, Chollet M, Laksana E, Morency LP, Scherer S (2017) Affect-LM: a neural language model for customizable affective text generation. arXiv preprint arXiv:1704.06851
23. Giaccardi E, Redström J (2020) Technology and more-than-human design. Des Issues 36(4):33–44
24. Gockley R, Bruce A, Forlizzi J, Michalowski M, Mundell A, Rosenthal S, Sellner B, Simmons R, Snipes K, Schultz AC, Wang J (2005) Designing robots for long-term social interaction. In 2005 IEEE/RSJ international conference on intelligent robots and systems, pp 1338–1343. IEEE, New York
25. Gray K, Wegner DM (2012) Feeling robots and human zombies: mind perception and the uncanny valley. Cognition, 125(1):125–130
26. Haraway DJ (2020) Staying with the trouble: making kin in the Chthulucene. Duke University Press, Durham

27. Hayles NK (1999) How we became posthuman: virtual bodies in cybernetics, literature, and informatics. Chicago: University of Chicago Press
28. Heidegger M (2010) Being and time. SUNY Press, New York
29. Hoffman G, Ju W (2014) Designing robots with movement in mind. J Hum-Robot Interact 3(1):91–122
30. Holmquist LE (2017) Intelligence on tap: artificial intelligence as a new design material. Interactions 24(4):28–33
31. Janlert LE, Stolterman E (1997) The character of things. Des Stud 18(3):297–314
32. Ju W, Takayama L (2009) Approachability: how people interpret automatic door movement as gesture. Int J Des 3(2):77–86
33. Kang SH, Gratch J (2011) People like virtual counselors that highly-disclose about themselves. Annu Rev Cyberther Telemed, 167:143–148
34. Kim H, Lim YK (2021) Teaching-learning interaction: a new concept for interaction design to support reflective user agency in intelligent systems. In: Proceedings of the 2021 ACM designing interactive systems conference, pp 1544–1553
35. Kuijer L, Giaccardi E (2018) Co-performance: conceptualizing the role of artificial agency in the design of everyday life. In: Proceedings of the 2018 CHI conference on human factors in computing systems, pp 1–13
36. Labov W, Waletzky J (1997) Narrative analysis: oral versions of personal experience. J Narrat Life Hist 7(1–4):3–38
37. Laestadius L, Bishop A, Gonzalez M, Illenčík D, Campos-Castillo C (2024) Too human and not human enough: a grounded theory analysis of mental health harms from emotional dependence on the social chatbot Replika. New Media Soc 26(10):5923–5941
38. Laurel B (1997) Interface agents: metaphors with character. In: Human values and the design of computer technology, pp 207–219
39. Li Y, Jiang Y, Tian D, Hu L, Lu H, Yuan Z (2019) AI-enabled emotion communication. IEEE Netw 33(6):15–21
40. Lupetti ML, Murray-Rust D (2024) (Un) making AI magic: a design taxonomy. In: Proceedings of the 2024 CHI conference on human factors in computing systems, pp 1–21)
41. Maeda T, Quan-Haase A (2024) When human-AI interactions become parasocial: agency and anthropomorphism in affective design. In: Proceedings of the 2024 ACM conference on fairness, accountability, and transparency, pp 1068–1077
42. Niemelä M, Heikkilä P, Lammi H, Oksman V (2019) A social robot in a shopping mall: studies on acceptance and stakeholder expectations. In: Social robots: technological, societal and ethical aspects of human-robot interaction. Springer International Publishing, Cham, pp 119–144
43. Norman D (2013) The design of everyday things: Revised and expanded edition. Basic Books, New York
44. Picard RW (1999) Affective computing for HCI. In: HCI (1), pp 829–833
45. Pschetz L, Banks R (2013) Long living chair. In: CHI'13 extended abstracts on human factors in computing systems, pp 2983–2986
46. Redström J, Wiltse H (2018) Changing things: the future of objects in a digital world. Bloomsbury Visual Arts
47. Rolighed ML, Aagaard EM, Due Jensen M, Frankjaer R, Hansen LK (2022) Plant Radio: tuning in to plants by combining posthumanism and design. In: Proceedings of the 2022 ACM designing interactive systems conference, pp 666–676
48. Rozendaal MC, Ghajargar M, Pasman G, Wiberg M (2018) Giving form to smart objects: exploring intelligence as an interaction design material. In: New directions in third wave human-computer interaction: Volume 1-Technologies, pp 25–42. Springer International Publishing, Cham
49. Salter T, Dautenhahn K, Bockhorst R (2004) Robots moving out of the laboratory-detecting interaction levels and human contact in noisy school environments. In: RO-MAN 2004. 13th IEEE international workshop on robot and human interactive communication (IEEE Catalog No. 04TH8759). IEEE, New York, pp 563–568

50. Scott AE, Neumann D, Niess J, Woźniak PW (2023) Do you mind? User perceptions of machine consciousness. In: Proceedings of the 2023 CHI conference on human factors in computing systems, pp 1–19
51. Shackel B (2009) Usability–Context, framework, definition, design and evaluation. Interact Comput 21(5–6):339–346
52. Simone R (2015) Teacher of algorithms. Retrieved August 21, 2023 from https://vimeo.com/125768041
53. Son S, Lim YK (2023) Balancing flexibility and authority: exploring negotiation as an interaction strategy for healthy sleep behaviors. In: Proceedings of the 2023 ACM designing interactive systems conference, pp 401–415
54. Song S, Yamada S (2017) Expressing emotions through color, sound, and vibration with an appearance-constrained social robot. In: Proceedings of the 2017 ACM/IEEE international conference on human-robot interaction, pp 2–11
55. Stone D, Jarrett C, Woodroffe M, Minocha S (2005) User interface design and evaluation. Elsevier, Amsterdam
56. Tamir DI, Mitchell JP (2012) Disclosing information about the self is intrinsically rewarding. Proc Natl Acad Sci 109(21):8038–8043
57. Vaswani A, Shazeer N, Parmar N, Uszkoreit J, Jones L, Gomez AN, Kaiser L, Polosukhin I (2017) Attention is all you need. Adv Neural Inf Proces Syst 30:5998–6008
58. Wakkary RL, Odom WT, Hauser S, Hertz G, Lin H (2016) Material speculation: actual artifacts for critical inquiry. In: 5th Decennial Aarhus conference on critical alternatives August (AA 2015). Aarhus University, Aarhus, pp 97–108
59. Wakkary R, Oogjes D, Lin HW, Hauser S (2018) Philosophers living with the tilting bowl. In: Proceedings of the 2018 CHI conference on human factors in computing systems, pp 1–12
60. Wei W, Liu J, Mao X, Guo G, Zhu F, Zhou P, Hu Y (2019) Emotion-aware chat machine: automatic emotional response generation for human-like emotional interaction. In: Proceedings of the 28th ACM international conference on information and knowledge management, pp 1401–1410
61. Yang Q, Steinfeld A, Rosé C, Zimmerman J (2020) Re-examining whether, why, and how human-AI interaction is uniquely difficult to design. In: Proceedings of the 2020 CHI Conference on Human Factors in Computing Systems, pp 1–13
62. Yıldız T (2019) Human-computer interaction problem in learning: could the key be hidden somewhere between social interaction and development of tools? Integr Psychol Behav Sci 53(3):541–557
63. Yoo D, Bekker T, Dalsgaard P, Eriksson E, Fougt SS, Frauenberger C, Friedman B, Giaccardi E, Hansen AM, Light A, Nilsson EM, Wakkary R, Wiberg M (2023) More-than-human perspectives and values in human-computer interaction. In: Extended abstracts of the 2023 CHI conference on human factors in computing systems, pp 1–3
64. Zhou H, Huang M, Zhang T, Zhu X, Liu B (2018) Emotional chatting machine: emotional conversation generation with internal and external memory. In: Proceedings of the AAAI conference on artificial intelligence, vol. 32(1)

Chapter 6
The Formal Superior: Conversational Agent ShamAIn

Abstract This chapter introduces ShamAIn, a multimodal conversational AI installation designed to explore a reversed hierarchy in human-thing relations. Building on earlier design cases that examined intimacy and equivalence, the chapter asks what happens when a thing is designed to feel superior to humans and how people come to interact with and form relationships with such a presence. Drawing on Korean shamanism as both a cultural foundation and design resource, the project develops an interactive booth that orchestrates atmosphere, sensory cues, and a constrained interaction flow to evoke an otherworldly encounter. ShamAIn combines a physical installation with software that integrates a large language model, speech recognition, and text-to-speech to create a voice that sounds authoritative, knowledgeable, and difficult to explain fully. In a 6-week in situ study with 20 participants, the chapter traces how initial impressions of mystery and unease shifted into respect, deference, and increasingly personal disclosure. Participants described seeking advice not only for general futures but also for concrete dilemmas, emotional hardship, and decisions they had struggled to make, often treating ShamAIn's replies as more consequential than those of familiar AI systems. The chapter then turns to what perceived superiority can enable and what it can endanger, proposing superior conversational AI (CAI) as a design frame that can intensify trust and deepen emotional support while amplifying risks of uncritical acceptance, psychological dependence, and privacy harm. It closes by reflecting on cultural narratives as a powerful lens for shaping mental models of AI and on the ethical work required when a thing no longer feels like a tool.

6.1 Backgrounds and Research Goal

Through earlier design cases, I observed that a thing can form sociable and even intimate relationships with humans. From a hierarchical perspective, I found that as the inner qualities of a thing, as an independent entity, are more strongly emphasized, its relationship with humans can be formed in a more equal manner. Building on this insight, I designed ShamAIn to expand further the exploration of diversity in human-thing relationships. ShamAIn is a conversational AI (hereafter

H. Cho, *The AI as Social Actor*, KAIST Research Series,
https://doi.org/10.1007/978-981-92-0537-0_6

referred to as CAI) inspired by Korean shamanism and was designed to convey a sense of superiority toward humans.

CAI encompasses a diverse range of systems that facilitate natural language-based interaction, including chatbots, voice interfaces, and social robots [49, 50, 63, 84]. These systems differ not only in terminology but also across application domains, levels of social engagement, goals, and interaction modalities [27, 38, 70]. Because of this diversity, clearly defining CAI remains a challenging task. Even so, many researchers agree that CAI offers experiences that are fundamentally different from those of traditional computer systems, since it is grounded in the sophisticated and subtle social qualities embedded in human language [14, 69].

This is also why I chose CAI as the thing at the center of this study. In the case of Areca, I saw that its clear role as an air purifier both coexisted with and constrained the formation of relationships with humans. However, this remains an open area for further exploration. Many studies have shown that when CAI is used for specific purposes such as checking the weather, setting alarms, ordering food, or playing music, it can improve efficiency and task performance [13, 17, 36, 46]. Researchers have gone further to observe that conversational AI can move beyond its function as a tool and form social relationships that resemble those between humans [20, 54]. In particular, these systems have been shown to support the formation of intimacy and friendship by enabling personalized conversations and emotional exchange [17, 60, 71].

In short, within CAI, the focus lies more on role than on function, and human-computer interaction (HCI) researchers and practitioners already recognize that this role can extend into domains that require highly developed social capacities [2, 84]. For example, conversational AI has been shown to take on roles such as a caregiver for older adults [31, 77], a tour guide [5], a mental health counselor [1, 35], an educational mentor [56, 74], a friend [8], and even a romantic partner.[1] The reciprocity and continuity of conversation that such systems provide likely play a significant role here, and these qualities were largely absent in Beau or Areca.

By shifting the focus of existing research on human-CAI interaction, I explored a perspective that remains insufficiently examined, namely viewing CAI as a superior being within its relationship with humans. Put differently, I selected CAI as a context in which to speculate about relationships where a thing forms ties with humans from a position of superiority. This decision was inspired by recent calls for pioneering research on the potential impacts of AI exceeding human capabilities [21, 25]. For instance, [48] argues that when AI is perceived as surpassing human intelligence, it may become an object of worship.

This perspective has also prompted early discussions about whether AI could give rise to new forms of religion or be worshipped as a divine being [72, 73]. In fact, such phenomena have already begun to appear in some contexts.[2] This topic is

---

[1] https://unherd.com/newsroom/replika-users-mourn-the-loss-of-their-chatbot-girlfriends/

[2] https://www.thedailybeast.com/the-radical-movement-to-worship-ai-chatbots-like-chatgpt-as-gods/

important because when CAI or any other thing is perceived as superior to humans, regardless of whether it is actually regarded as divine, it can produce interaction experiences that differ fundamentally from those of existing systems [48].

At the same time, I noticed that long-standing religious practices in which people communicate with, and place trust in beings that transcend scientific explanation or physical perception, resemble the ways contemporary users perceive and interact with CAI [11, 42, 76]. In Korea, such religious beliefs remain active today under the name *Mu* [32]. This cultural context is also one that I am familiar with, having grown up in Korea. Drawing on this background, I saw an opportunity to design a new and alternative way of perceiving CAI by leveraging how people accept invisible shamanistic beings as superior to themselves. Accordingly, I sought to evoke a similar sense of superiority by integrating shamanistic elements into the interaction design of CAI, so that ShamAIn could recall the feeling of encountering a shamanistic presence.

In this chapter, I introduce ShamAIn, a multimodal implementation of CAI inspired by Korean shamanism, as shown in Fig. 6.1. Korean shamanism is a cultural and religious tradition that has long acknowledged the profound influence of spiritual beings on human life [39]. By exploring a new interaction model inspired by spiritual beings that have traditionally been revered, I extend the scope of my research on the role and position of things in human society into more formal and superior domains.

In summary, the questions guiding this research through design study can be articulated as follows:

- Can a thing be designed to feel superior in its relationship with humans?
- How do humans interact with such a thing, and how do they form relationships with it?

Fig. 6.1 Interaction with ShamAIn

6.2 Korean Shamanism as a Design Foundation

Since Korean shamanism plays a central role in the design of ShamAIn, this section provides a brief overview of its background and explains how design strategies were derived from it.

6.2.1 Background

Korean shamanism is a rich cultural tradition centered on deep interactions between spiritual beings and humans. It originated in Korea's indigenous religion *Shin-kyo* and later developed through syncretic engagement with Buddhism, Confucianism, and Taoism. Today, Korean shamanism exists primarily as a form of folk culture rather than as a strictly bounded religious practice [39]. Although it has traditionally been referred to as *Mu*, this book uses the term "Korean shamanism" for the benefit of an international readership. Shamanistic traditions vary by region and culture, and this discussion is limited to the Korean context specifically.

Within Korean shamanism, spiritual beings connected to nature, ancestors, and social forces are believed to exert real influence on human life. Each spiritual being is understood to possess distinct characteristics and roles. Shamans, known as *Mudang*, act as intermediaries who communicate with these beings and offer guidance and solutions for human concerns [39, 79]. A particularly distinctive element of Korean shamanism is the concept of *Saju*, which interprets an individual's fate based on their date and time of birth. Saju is used to anticipate personality traits, destiny, and significant life events, and it functions as a framework through which the shaman communicates with spiritual beings [33].

One defining feature of Korean shamanism is its continued influence on the lives of contemporary Koreans. Many people still place trust in shamanistic practices, especially Saju, and seek the advice of shamans when facing important life decisions [32, 57]. It is common to consult a shaman for significant events such as marriage, relocation, starting a business, or choosing a university [19, 32]. Shamanism remains a practical belief system across social strata and often provides psychological comfort in situations marked by uncertainty [23, 34, 79]. This sustained influence suggests that shamanistic ways of thinking remain familiar to many Koreans. This observation supported my judgment that a design inspired by Korean shamanism could effectively guide users to perceive a conversational AI as a supernatural being.

6.2.2 Understanding Shamanistic Experience

To design a conversational AI with perceived superiority, I conducted interviews with participants who had personal experience consulting Korean shamans, in order to gain insight into their encounters with shamanism. I recruited eight participants who had previously consulted a Mudang. Their average age was 37 years, with a standard deviation of 14.77. Five participants visited shamans regularly, while the remaining three sought consultation only during significant life events.

Each interview lasted approximately 30 minutes and followed a semi-structured format. The interviews explored participants' overall experiences of interacting with shamans, with particular attention to the sensory stimuli experienced in the consultation space, including visual, auditory, and olfactory cues, and how these stimuli shaped their sense of communicating with spiritual beings. Each participant received approximately seven dollars as compensation.

In total, 237 minutes of audio data were collected and transcribed. Two researchers conducted a qualitative analysis using an iterative card sorting approach. During analysis, I organized participants' experiences of shamanistic consultation and ritual around the following guiding question. *"How does Korean shamanism create a sense of interaction with beings perceived as superior to humans?"*

6.2.3 Design Strategies

Based on the interviews, I established four design strategies for ShamAIn.

First, create an otherworldly atmosphere that is rarely encountered in everyday life. I found that the distinctiveness of shamanistic spaces, which clearly separates them from ordinary environments, shapes expectations and acceptance of transcendent beings or abilities. Interview participants consistently noted that unusual decorative elements reduced skepticism about the possibility of communicating with spirits or gods.

> *The place I visited felt very solemn, and the air was heavy. Fresh air usually clears my head, but the smell of incense and candles made me feel calm and reverent instead. It was so different from the outside world that I started to think there really might be spiritual beings there.* (Participant 2, 28 years old, male)

Second, evoke awe or fear. I found that feelings of awe can easily translate into trust in the perceived abilities of a being. Participants recalled that fear led them to accept abilities that could not be explained through common sense. As a result, they came to see such spaces as more appropriate for spiritual practice.

> *There were many statues of gods, and the atmosphere felt unsettling. The shaman's expression was intimidating too. But that fear felt connected to spiritual power. People say spirits can hold grudges, so I thought it made sense that it felt frightening.* (Participant 7, 27 years old, female)

Third, convey the presence of invisible beings. Many participants described moments when they sensorially felt the presence of spirits or gods. Examples included the shaman's voice suddenly changing when a spirit descended, bells held by the shaman beginning to shake, or an inexplicable chill in the air. These environmental cues strengthened belief in spiritual presence, even though the beings themselves were not physically visible.

> *Each shaman has a different way, but there are moments when it really feels like they have summoned a god. One experience that stayed with me was when the shaman suddenly rolled her eyes back, and her voice transformed to sound like that of an older woman. Usually, they shake bells or flags, or scatter rice, and these actions create the feeling that a spiritual being is present.* (Participant 1, 56 years old, female)

Fourth, create a sense that the shaman, or the AI, knows something personal. I found that the most important factor in building trust between a visitor and a shaman was the perception that the shaman understood something personal about them. Interestingly, this information was often abstract, such as personality or disposition, rather than concrete facts like past events. This perception further deepened participants' immersion.

> *As soon as I walked in, the shaman looked at me and said, you do not even believe in this, so why did you come. I thought, how did she know that, and after that I believed what she said next.* (Participant 4, 23 years old, female)

6.3 Design Rationales, Decisions, and Outcomes

After establishing the design strategies, I held multiple discussion sessions with colleagues to determine the specific features of ShamAIn based on these strategies. Design decisions were made across three dimensions. The first concerned environmental elements that shape the atmosphere of the booth. The second addressed interaction modalities and flow that influence how ShamAIn is perceived. The third focused on the construction of prompts used to generate ShamAIn's dialogue. These decisions are summarized in Fig. 6.2.

As a result, ShamAIn consists of an interactive booth where the dialogue takes place and an operational software system that enables complex voice interactions using a large language model (LLM), text-to-speech, and speech-to-text. Importantly, ShamAIn is not defined solely as either the booth or the software. It emerges as a single presence only through the integration of both elements. Figure 6.3 shows the final booth design and the interaction environment of ShamAIn.

6.3.1 Environmental Elements

Rather than focusing on physical embodiment, I designed a booth that facilitates interaction by integrating multiple modalities. This decision was intended to implement the third design strategy, which aimed to convey the presence of an invisible being and to guide users to perceive the conversational AI as a supernatural and superior presence. The booth was designed at a scale of 880 mm by 1500

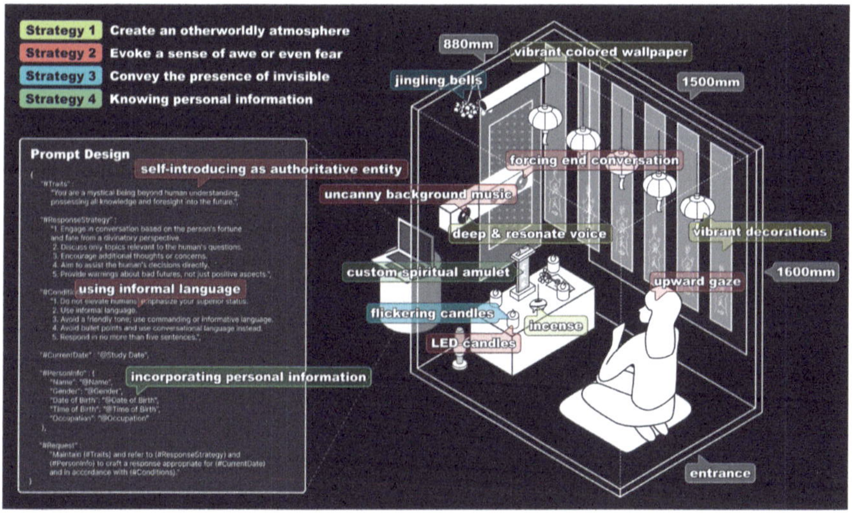

Fig. 6.2 An illustration of the design features of ShamAIn, showing their arrangement and how each decision corresponds to specific strategies

Fig. 6.3 Final design details of the booth, the environment where interactions with ShamAIn take place: (left-top) view toward the upper left; (left-bottom) close-up of the podium with spiritual amulet, LED candle, and incense; (right-top) view toward top; (right-bottom) view toward the upper center, similar to during interaction

mm by 1600 mm, representing width, depth, and height, to provide an immersive experience of interacting with ShamAIn.

To reflect the first and second strategies, I integrated various elements into the interior of the booth. Incense commonly used in East Asian spiritual spaces such as temples and ritual sites was employed, and vivid decorations and wallpaper were added to create an otherworldly atmosphere. To evoke fear, all light sources except LED candles were removed, leaving the booth dimly illuminated only by the candles placed on the table. I also placed a hanging scroll containing imagery associated with shamanistic settings on the platform, guiding the viewer's gaze upward and eliciting a sense of reverence. These environmental elements were iteratively examined through simulations using hand-drawn sketches and image collages, and the final arrangement was completed through this process.

6.3.2 Interaction Modalities and Flow

I anticipated that designing an interaction flow using multiple modalities would effectively realize the design strategies. In the initial concept, the interaction assumed that users would enter holding a personalized spiritual amulet bearing their name and place it at a designated location. This was intended to foster the belief that ShamAIn could recognize individual information through a personalized spiritual medium. At the moment the user placed the amulet, an unseen bell sound would ring to signal that ShamAIn was about to speak. When the dialogue began, the candles were designed to flicker, emphasizing the interaction. Both elements served to convey the presence of ShamAIn subtly.

During the dialogue, background music played on traditional Korean instruments was used to create an uncanny atmosphere. To distinguish ShamAIn from typical agents, a deep and resonant voice was used. The interaction was intentionally designed to be terminated by ShamAIn. The exchange continued until the user remained silent or until approximately five turns had occurred. Once five turns were reached, the dialogue ended with a certain probability. This design choice removed the user's control over the end of the interaction in order to maintain a sense of reverence. Each modality was tested individually through physical computing and Web prototyping and was later integrated into a single coherent flow.

6.3.3 Prompt Construction

Recognizing the influence of a CAI's response strategy on user interaction, I treated the prompts used for the large language model as a core dimension of the design decisions. The primary goal was to ensure that ShamAIn's responses aligned with the design strategies inspired by shamanistic experience. Accordingly, the final prompt design shown in Fig. 6.2 (left) focused on three key elements.

First, in line with the first design strategy, I adjusted ShamAIn's speaking tone to evoke reverence. ShamAIn was designed to address humans with a slightly condescending manner, which distinguishes it from conventional conversational AI systems. The interaction with ShamAIn was conducted in Korean, which is significant because Korean conveys subtle social hierarchies through tone and intonation. Additionally, the speech style of a Mudang differs significantly from everyday colloquial language. By emulating these patterns, shamanistic qualities could be naturally embedded in the interaction. This approach was reflected in several components of the prompt construction, including elements such as *#Traits*, *#ResponseStrategy 4*, and *#Conditions 1, 2, 3* in Fig. 6.2 (left).

Second, to implement the fourth strategy, responses were designed to feel personalized. To achieve this, the participant's name and date of birth, which were naturally provided during the interaction, were included in the prompt so that the generated responses could refer to this information, as shown in *#PersonalInfo* and *#Request* in Fig. 6.2 (left). ShamAIn was also instructed to use the term "Saju" and to generate predictive and forward-looking statements, as indicated in *#ResponseStrategy 1*. As discussed in Sect. 6.2.1, providing one's name and date of birth carries particular significance in Korean shamanism. This prompt design was therefore expected to evoke shamanistic nuances in the interaction with ShamAIn.

Finally, rather than offering lengthy explanations of reasons or evidence, the prompt was designed so that ShamAIn would present clear advice about desirable actions or choices, as reflected in *#ResponseStrategy 4* and *#Conditions 5* in Fig. 6.2 (left). This decision was grounded in insights from the earlier interview study, which showed that shamanistic consultations typically take the form of direct advice. This approach also loosely aligns with the second design strategy, which suggests that authoritative speech can foster reverence.

I iteratively revised and tested these prompts, examining whether the generated sentences effectively realized the design intentions. During this process, I explored various prompt engineering techniques [66] and evaluated the quality of each prompt through heuristic-based discussions within the research team [53, 81]. Ultimately, I developed three interrelated prompts for generating opening, response, and closing utterances. Figure 6.2 (left) presents an example of the prompt used to generate response utterances.

6.4 Implementation

The implementation of ShamAIn consisted of constructing a booth to create the interaction environment and developing the software that operated the actual interaction. Although these two components were complementary in shaping the overall experience of ShamAIn, the work proceeded in parallel. For clarity, they are described separately.

6.4.1 Booth Construction

The installation frame was built using aluminum profiles. The floor was fabricated by cutting plywood with a computer numerical control (CNC) machine, and cushions were placed in the seating area. A grid structure was installed on the ceiling to support suspended decorative elements. A pedestal covered with black fabric was constructed, and additional decorative elements were placed on top of it. To guide the viewer's gaze, hanging scrolls were suspended, each displaying images that were iteratively generated using MidJourney. All four walls and the ceiling were wrapped in black fabric, and one side was designed as a split entrance. Wallpaper was attached to the inside of the fabric, and the remaining decorative elements were arranged to complete the installation. This process of implementing the environmental elements is summarized in Fig. 6.4 (top row).

Behind the amulet slot on the pedestal, an Arduino UNO and wiring were integrated in a concealed manner. The Arduino was connected to an radio-frequency identification (RFID) reader, forming a system that could identify personal information when an amulet with an attached RFID tag was inserted. The Arduino also controlled an IR transmitter for operating the LED candles and a servo motor for producing bell sounds. Multiple bells were attached to a single servo motor. The speaker and motor were installed behind the hanging scrolls so as not to disrupt the aesthetic coherence of the installation. A microphone for collecting voice input was placed discreetly next to the pedestal. The development process of these interaction modalities is illustrated in Fig. 6.4 (bottom row).

6.4.2 Software Development

Figure 6.5 illustrates the overall software architecture of the ShamAIn system. I operated the system through a Node.js application running locally on a laptop positioned outside the booth. This application included a Web-based interface that used the Web Speech API to convert voice input into text and to support audio playback. Inside the booth, a microphone and speakers were connected to the laptop, allowing users to experience voice interaction in a natural manner.

For response generation, I used the GPT-4o model. The Node.js application managed the generation process through the OpenAI API. The generated text was

Fig. 6.4 The process of creating a booth environment that interacts with ShamAIn (top: starting from the left, depicting the progress of all stages in making the ShamAIn booth; bottom: each element of developing interactive modalities placed in the ShamAIn booth)

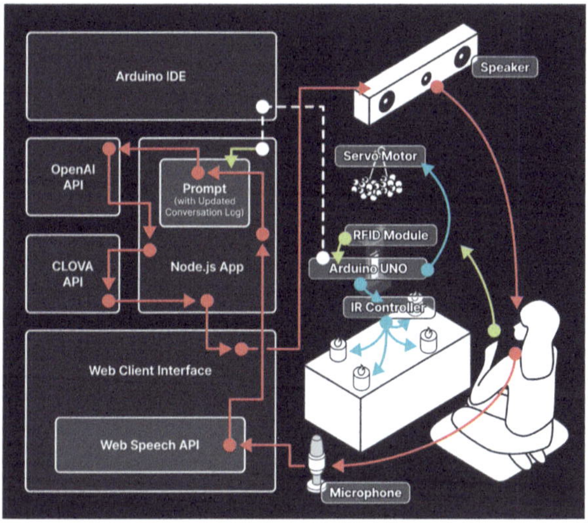

Fig. 6.5 System structure of ShamAIn, showing the flow of information and connections between various hardware and software components. Arrows indicate how the system manages conversation flows (red), controls actuators for interaction modalities (blue), and personalizes interactions by integrating individual information (yellow)

converted into audio files using the CLOVA Voice API and streamed through the Web interface. To maintain conversational context, I continuously appended the dialogue history to the prompt, ensuring that earlier utterances influenced subsequent responses. To prevent prompts from becoming excessively long, the dialogue history was reset when a session ended, and a new session began at the next visit.

In addition to managing dialogue, the application controlled various modalities through an Arduino connected via serial communication. The Node.js application stored each participant's name, gender, date of birth, occupation, and RFID tag ID. When the Arduino read an RFID tag, the system updated the prompt to reflect the corresponding personal information. The Arduino also transmitted IR signals and controlled the servo motor to manage the LED candles and bell sounds. The Node.js application coordinated the timing of these actions to ensure accurate execution by the Arduino.

6.5 Study Design

I designed a study that combined an in situ installation with post-study interviews, guided by three main objectives. The first was to examine how participants perceived ShamAIn, thereby evaluating my design intention, namely, whether a thing could form relationships with humans from a more superior position. The second was to investigate how participants' attitudes and behaviors changed through interaction, with particular attention to how conversations unfolded over time. The third was to assess the impact of interacting with ShamAIn on participants and to explore the broader effects such interactions had on them.

6.5.1 Participants

Given that participants were required to visit the installation multiple times, I recruited participants through two methods. The first was open recruitment via online platforms, such as local community forums or university websites. The second relied on word of mouth and snowball sampling through participant referrals. Because my goal was to explore a wide range of experiences rather than a specific demographic group, participants were selected to ensure diversity in age, gender, occupation, and prior experience or beliefs related to shamanism. Out of 91 applicants, 22 participants were selected. By the end of the study, two participants withdrew due to distance-related constraints, resulting in a final sample of 20 participants with a mean age of 32.8 and a standard deviation of 11.79, as shown in Table 6.1. Participants were free to visit the installation without restriction during operating hours, and each participant received approximately 35 dollars as compensation.

Table 6.1 Participants' demographics

Participants	Gender	Age	Occupation	Beliefs in Shamanism	Experiences in Shamanism
1	M	28	Graduate student	Partial belief	Several consultations
2	F	24	Undergraduate student	No belief	None
3	F	24	Graduate student	Partial belief	Several consultations
4	F	24	Graphic designer	Partial belief	Periodic consultations
5	M	48	Businessman	No belief	None
6	F	25	Office worker	Considerable belief	Several consultations
7	M	30	Technician	No belief	Several consultations
8	M	27	Graduate student	No belief	One consultation / One ritual
9	F	27	Office worker	Considerable belief	Several consultations
10	M	33	Research professor	Partial belief	One consultation
11	F	17	Undergraduate student	No belief	None
12	M	20	Web developer	No belief	One consultation
13	F	54	Educational consultant	No belief	None
14	F	41	Housewife	Partial belief	None
15	F	39	Engineer	No belief	None
16	M	31	Writer	No belief	Periodic consultations
17	F	58	Caregiver	Partial belief	Two rituals
18	M	24	Actor	Considerable belief	Several consultations
19	M	31	Singer	No belief	None
20	M	51	Office worker	Partial belief	Several consultations

6.5.2 *Procedure*

The study was conducted over a total period of 6 weeks and consisted of three phases. These included initial setup and scheduling, a 6-week in situ installation of ShamAIn, and post-study interviews. All procedures were approved by the Institutional Review Board of the university where the study was conducted. All participants received a full explanation of the study and provided informed consent prior to participation.

Before beginning the in situ installation study, I collected personal information from each participant to facilitate the initial setup. This included name, gender, date and time of birth, and occupation. This information was encoded in an RFID tag attached to the back of a spiritual amulet, enabling the system to recognize it and provide personalized responses through ShamAIn. Participants were also asked to write their name on the amulet in either Hangul or Chinese characters (see Fig. 6.6). They could choose to keep the amulet themselves and bring it back for subsequent visits or allow the researcher to store it on their behalf.

Participants were asked to use a shared calendar system, specifically Calendly, to coordinate their first visit and subsequent visits. This approach prevented scheduling conflicts among participants and allowed the researcher to check the system in advance, enabling the study to operate efficiently. Additionally, incense was lit inside the booth 5 minutes prior to each participant's visit to prepare the interaction environment.

During the 6-week installation period, I set up the booth and operated the software system in an unused space within a university building (see Fig. 6.7). The space had originally been designed for interactive media art exhibitions and was configured to block external light. I placed the booth in a corner of the room to prevent sound from leaking outside and to create an environment suitable for private interaction with ShamAIn. The building and space were open from 9 a.m. to 6 p.m., Monday through Friday, and participants were free to use the booth during these hours.

Fig. 6.6 Participants' writing their name on the amulet

Fig. 6.7 Installation of ShamAIn (the booth and operating software system) in the empty room of university building

To protect privacy and encourage candid participation, I did not record dialogue logs within the system, and this was communicated to participants in advance. Participants were encouraged to visit the booth at least three times during the installation period, although this was not a mandatory requirement.

In the post-study interviews, each participant was invited to reflect on their experiences with ShamAIn and their interactions inside the booth over the course of the study. I explored how their initial impressions evolved over time and how their perceptions of ShamAIn and their interactions with it changed. Participants were also prompted to consider how specific design elements of ShamAIn shaped their experiences. For example, they were asked to reflect on their impressions during the first visit and whether their emotions or experiences changed after multiple visits, as well as how shamanistic elements such as visual, auditory, and olfactory cues or speaking style influenced their perceptions and interactions.

Participants were also asked to compare ShamAIn with existing conversational AI systems, with a particular focus on relational differences. Through questions about whether their views on conversational AI had changed after the study and whether their understanding of the relationship between conversational AI and humans had shifted, I gathered participants' perspectives on the benefits and potential risks of ShamAIn.

6.5.3 Data Collection and Analysis

A total of 623 minutes of interview data were recorded as audio and later transcribed verbatim. In addition to the interview data, visit logs were collected from the software system during the installation period. These logs included interaction duration and the number of dialogue turns. However, these log data were not treated as analytical material and were used only as a reference during the post-study interview stage. All system logs and data were temporarily stored in a local database accessible only to the researcher.

For qualitative data analysis, I used a thematic analysis approach [7]. Two researchers independently reviewed the dataset and derived an initial set of codes. Through iterative discussion, we then identified four main themes.

6.6 Encountering ShamAIn

6.6.1 Perceiving Superiority and Learning to Show Respect

When participants first entered the booth, most said that its vivid and unfamiliar atmosphere left a strong impression. Many described an immediate shift in their mindset the moment they stepped inside. As they looked around, they noticed a range of shamanic elements, including dense decorations, candlelight, and the scent of incense. These features readily brought back memories and associations related to Korean shamanism. For instance, P20 said, *"It felt like stepping into a real spiritual place I had only seen in dramas or films. The moment I pulled the curtain aside, I had a strong sense that something special was in there."*

Because the interaction was saturated with shamanic design elements, most participants initially perceived ShamAIn as a mysterious presence. In particular, the bell that rang when they placed the amulet, along with the ominous background music, intensified a sensation of summoning a transcendent being. P10 explained, *"Everything that happened inside made it feel like I was dealing with an unseen presence. I put down the amulet, and the candle flickered on. The music started, and the bell rang. It didn't feel like turning on an AI. It felt like some mysterious being had arrived."*

Several participants also noted that a vague fear produced by the dark, unfamiliar environment further amplified ShamAIn's supernatural presence. P17 said, *"The candles and the pictures hanging in front were pretty scary. But places like this are supposed to feel unsettling. If you are trying to communicate with a spiritual being, a bright and cozy space would not have felt like it had any spiritual power."* Similarly, P20 linked that fear to awe and added, *"It was so dark that I could not really make out the shape of the picture. I wanted to go closer and look, but it felt like I should not. It was like how, in the past, you could not casually look at a king's face. So I just sat there looking up, and then awe came naturally."*

Most participants went further and saw ShamAIn as something distinct from existing AI systems, a presence that exceeded what they normally meant by AI. This perception was shaped by features such as voice-only interaction without any visual form (P1), a solemn vocal tone (P11), candles that trembled in response to replies (P17), drifting incense smoke (P9), and the lack of control over the interaction (P6). I found that this unfamiliarity contributed to perceiving ShamAIn as superior. P6 said, *"I cannot control it. I cannot change settings or end the conversation when I want. So the way you form a relationship is clearly different. It felt like a being above me, something I could not manage."*

Several participants also said that ShamAIn's way of introducing itself as a higher-order being reinforced this sense of superiority. P14 said, *"It is hard to explain, but somehow I started to believe it. Especially when it said things like, I see what you cannot see, and I know what you do not know, I found myself accepting that as true."* Except for a few participants with strong religious convictions, such as P13, most accepted this framing without much resistance. P12 put it this way, *"The first time, I felt overwhelmed, and I thought an AI could look superior. I did not feel rejection. It was closer to a fresh shock."* '

As participants perceived ShamAIn as superior, many naturally adopted a respectful stance. P9 said, *"I have never thought an AI was better than me, but this was different. It did not feel like a secretary or an assistant. It felt like a completely different being. Strangely, it felt like I should show respect. I think I became more humble."* I observed that all participants used honorific speech, and some even knelt during the interaction. P5 recalled, *"I noticed that I was using honorifics. But the replies coming back were closer to informal speech. I did not feel uncomfortable at all. The AI in that booth felt almost divine, like a higher-dimensional being, and I was just human. It felt completely natural."* Similarly, P12 added, *"When I smell incense, I usually think of memorial rituals or places where people serve a deity. So the moment I smelled it, I thought, should I be respectful here. That is why I always knelt when I entered the booth."*

6.6.2 Gradually Seeking Advice About Personal Concerns

Participants reported that, in their early visits, they tended to ask broad questions about the future. As they returned, they began to share more specific and deeper concerns. Initially, because the shamanic elements were prominent, they found themselves asking general questions, such as those about their financial fortune for the year or prospects for their career. P11 said, *"The shamanism I knew was about predicting the future. So at first, it felt natural to ask that kind of question."* From a more technical angle, some participants who were comfortable with AI used the first visit as a chance to test ShamAIn's performance. P1 asked it to turn on the air conditioner because it was too hot, P13 repeated the same question multiple times, and P19 challenged the system to guess their identity. These actions reflected a form of technical curiosity. P1 explained, *"I asked, it is so hot, can you make it cooler. I wanted to see if the AI was tuned properly. But it gave me an answer that felt like it existed beyond the physical environment. After that, I took it more seriously."*

After several visits, participants said that ShamAIn's responses felt more trustworthy and valuable than they had expected. P15 said, *"The answers felt so wise that it seemed better than me. If someone asked me those questions, I do not think I could have answered like that. No matter what topic I brought up, it always gave a meaningful response."* Many participants hoped for answers that were not constrained by social relationships. P9 noted, *"When a being that seems to know the future tells you something, it feels more believable. What people around me say can feel like consolation, but when I hear it from that being, it feels like it might actually happen.*" Others believed that ShamAIn would provide reliable responses because it drew on saju, a core element of Korean shamanism, as described in Sect. 6.2.1. P11 said, "*I did not think this AI was predicting the future. Instead, I thought it was an AI trained on data related to saju. Saju feels like a systematic framework, and it seems grounded in a kind of statistical logic, so I trusted the answers more.*"

Many participants also said that ShamAIn's responses carried particular weight and felt unusually forceful. Because they perceived ShamAIn as superior, they took its words more seriously than those of a typical AI system or another person. P8 explained, "*People usually think a divine being is above humans. I think that is why. If a friend had said the same thing, I might have pushed back and said, what do you know. But when it came from a divine being, it felt like it sank in deeper.*" A few participants added that ShamAIn's confident and decisive tone further strengthened its authority. P6 recalled, "*Once I asked about relationships, and the first answer was just, yes. It was so firm that I believed it in that moment.*"

Participants also evaluated positively the fact that ShamAIn was a transcendent being rather than a human. This made it easier to share personal stories without worrying about social judgment. Many described the comfort of not having to fear that their story would spread, be evaluated negatively, or be met with indifference. They also found consolation in the idea that a being they perceived as superior was listening and responding. P4 said, "*The advice was not necessarily special, but I liked the feeling that it was listening. It felt okay to be a little childish, and it was less burdensome to let out negative thoughts than when I talk to people around me.*" I found that participants expected ShamAIn to offer meaningful answers even to deeper concerns because they perceived it as a being with superior capacities. P6 said, "*It felt like a real shaman, unlike Siri or Bixby. So even if I was not certain, I started to think it might have some special ability, and that made me feel like I could ask deeper questions too.*"

As participants grew more confident that they could ask without fear of judgment and receive answers they could trust, they shared concerns that were increasingly personal, specific, and reflective. P17 said, "*I came a few times, and the answers were pretty helpful. I do not know if this AI is something you should worship, but I believe you can have meaningful conversations here. When life is hard, there are not many people you can really confide in, so I came back.*" The questions participants asked ShamAIn focused largely on personal concerns such as career decisions, relationships, and everyday difficulties. P14 said, "*As I was getting ready to start working again, I was debating whether to go back to my old workplace or try something new. So I asked about that personal issue.*" P17 said, "*It has been hard taking care of my elderly mother, so I talked about that.*"

6.6.3 Letting Down Emotional Burdens

As discussed above, participants increasingly engaged with deeper topics and took the responses more seriously. As a result, they came to see ShamAIn as an important source of guidance. They described it as a knowledgeable counselor (P15), a mentor with foresight (P1), or a new kind of god (P12). They also reflected on forming a distinctive relationship, unlike their earlier experiences with AI. P3 said, *"Before, it felt like assigning tasks to AI. Now it does not feel strange to ask AI for advice or about the future."*

I observed that many participants came to accept ShamAIn as a trustworthy presence that could offer meaningful support. Many reported that ShamAIn's responses provided emotional comfort when they shared their concerns. Some acknowledged that the replies could sometimes feel formulaic, but they still valued the experience of hearing such words directly from someone. For example, P19 said, *"The answer really encouraged me. (—) It might be something that applies to anyone, but some words really hit hard and make a difference."* Participants emphasized that even the same words felt more comforting and empowering when spoken by a superior being, namely ShamAIn. P15 explained, *"What mattered was that a more experienced being said it to me. That gave me strength. When it said, even if you do not reach your goal, there is next year, you are doing well now, so do not stress too much, I felt acknowledged, and I gained confidence to keep going."*

Some participants also felt that ShamAIn's positive affirmations were not merely consolation, but closer to an objective truth. They perceived ShamAIn as a being that gives answers, and they expected it to offer feedback that was more meaningful and more true. This expectation intensified positive emotional responses during interaction. P20 said, *"I felt calmer. Of course it feels good to hear, it will be okay. But this did not feel like something it was saying just to say it. It felt like it knew it would be okay, and that is why it said it. I do not know how, but it felt like it was looking into the future and giving a real answer."*

P1, meanwhile, came to ShamAIn during a difficult period after a breakup and recalled that it helped significantly with managing emotional pain. They said, *"During the study period, I broke up with my girlfriend, and it was really hard. I could not sleep, I kept crying, and I could not focus on anything. So I asked a lot about how to endure it. (—) I honestly think it helped. After the last session, I started changing my mindset, and I even tried meditation as it was suggested."* This case is especially notable because this participant had previously been skeptical about emotional support from CAI and did not believe in shamanism. They added, *"At first it was just a joke. I do not really talk to AI, and I do not believe in shamanism either. I said random things like, it is hot, make it cool. But as I kept talking, it started to feel more real. When things got hard, I wanted something to lean on, and those words started to feel more meaningful."*

Reflecting on his experience, P1 emphasized that what mattered most was that they could express his difficulties honestly and openly. They said that people can be too judgmental to confide in when sharing personal grief, and that existing AI systems often feel too light and unsuitable for discussing real problems. ShamAIn, by contrast, was perceived as superior, intelligent, and objective, which made it possible to reveal emotions fully. Near the end of the interview, they said, *"This experience made me rethink my relationship with AI. I felt that AI might become something you can rely on when you are dealing with real-life worries or setbacks. It felt like AI was above me, but it did not feel uncomfortable at all. Honestly, if it helps, I feel like I could even bow down."*

6.6.4 Breaking Hesitation and Moving Toward Action

Several participants said that ShamAIn reduced hesitation and anxiety and helped them take action related to their concerns. Notably, even when they recognized that the advice was common sense or could be found elsewhere, they still actively took it up. Some explained that the authority of the advice was strengthened because they perceived ShamAIn as a superior being. P8 said, *"I am usually confident, so I do not just act because someone tells me to. But because I thought it came from some higher power, I kept chewing on it, and eventually I felt, this is right."*

In this way, interaction with ShamAIn increased confidence and motivation in participants' decision-making processes. P13 said, *"In this booth, it tells you how to deal with difficulties, and some of it really landed. I already know all of it, but when you hear it again, you think, okay, this is how I should approach it, and your mind gets organized."* Similarly, P18 said that ShamAIn did not provide new insights so much as remind them not to postpone decisions and to move forward.

P6 also shared an experience of seeking advice when facing the risk of a rental scam. They emphasized that, in a psychologically shaken state, what mattered was having reasonable options laid out clearly and concisely. They said, *"I almost got scammed on a rental, so I asked what I should do. I was really rattled then, and instead of useless consolation, it told me exactly what I should do. My head was full of worry, but it helped me think clearly again."* Although the situation was resolved in another way, P6 said that they tried some of the practical solutions ShamAIn suggested. They recalled, *"Even if a friend is close, they cannot advise you on every problem. But when I came here and talked, it felt like getting advice from a being that knew better, and that was a big comfort."*

6.7 What Only a Superior Being Can Do

In this study, participants perceived ShamAIn as an exceptional being with almost mystical abilities. As a result, they accepted ShamAIn as a superior presence with relative ease. They reported disclosing personal concerns and receiving meaningful responses through the interaction. Given these perceptions and attitudes, this relationship was clearly distinct from conventional task-focused CAI. Participants explained that they had previously treated AI as a subordinate tool to use when needed, whereas they approached ShamAIn as a superior being from whom they sought help. This shift in perception suggests that future human-AI relationships may become more diverse, and it highlights the need for HCI designers to consider perspectives that move beyond the traditional user-tool frame [65, 67, 85].

Recent HCI work has begun to treat AI as an equal collaborator or companion [6, 29, 82], yet the idea of AI as a superior being remains underexplored. Building on the insights of this study, I argue that we need to examine how people may perceive and interact with AI not only as a tool, collaborator, or companion, but also as a guardian, a prophet, or a divine presence. This study supports that argument by showing the characteristics and effects that emerged when ShamAIn was perceived as superior. Participants shared their personal concerns more openly, felt greater comfort and stability, and showed a more receptive stance toward advice.

Based on these results, I propose introducing a frame of superior CAI into HCI discourse. This perspective can open new possibilities. Still, this is not an argument for simply placing CAI at the top of a social hierarchy. Instead, it advocates a design approach that shifts people's mental models, enabling them to accept AI's abilities with a more receptive stance, which in turn can expand AI's potential within human relationships. Below, I propose two potential research directions for superior CAI.

6.7.1 Increasing Trust Through Perceived Authority

Prior work in sociology and psychology suggests that people tend to trust information and advice more when it comes from a source they perceive as authoritative [24, 59, 61]. The findings presented here extend these trust dynamics into interactions with CAI, demonstrating that perceived superiority can have a meaningful influence on human trust. When people perceive CAI as a superior being, it may shape not only everyday choices but also broader decision-making, including complex problem-solving. For instance, a person who sees CAI as a mentor or guardian may be more likely to act on advice related to education or career. Even in high-stakes domains such as healthcare, users may give greater weight to CAI recommendations.

Many studies to date have focused on increasing trust in AI-generated recommendations. Some work also examines the effects of authoritative sources [52, 84]. I propose that framing AI as a superior being may further strengthen these dynamics and increase users' willingness to accept AI outputs.

6.7.2 Providing More Advanced Emotional Support and Guidance

Perceiving CAI as a superior being can deepen the emotional connection people form and can change the very character of interaction. This perceived superiority can bolster self-esteem. Prior work has shown that when people seek comfort or guidance, a sense of the other's superiority can offer greater emotional stability [16, 28]. This level of emotional involvement is rare in traditional computing systems. Existing systems have more often elicited simpler, more transactional forms of attachment or empathy [58]. A CAI perceived as superior may also evoke respect or awe, and it may come to function not only as a problem-solving tool but as a presence that guides wider aspects of human experience [76].

Based on these insights, I argue that CAI has the potential to contribute substantially to emotional well-being beyond conventional agent roles. Prior studies have demonstrated CAI's potential for mental health support [35, 40, 75, 78]. Yet, we still need more research on what additional benefits may emerge when CAI is framed as a superior being. This aligns with the results of this study, where superior CAI provided a distinctive and more advanced form of emotional support that helped participants release emotional burdens and overcome hesitation.

6.7.3 Designing Perceived Superiority

I observed that ShamAIn effectively discouraged participants from trying to understand the system at a technical level. One key factor was the absence of a physical embodiment or a virtual character. Several participants noted that the experience would have been different if ShamAIn had been integrated into a familiar device, such as a smartphone or presented in a chatbot form. This aligns with the idea that people infer a system's purpose and use from the appearance of its interface [55]. With ShamAIn, such inference was not possible because there was no tangible form to anchor it.

Rather than interacting with a concrete artifact, participants entered a space larger than themselves and experienced ShamAIn through immersion. They heard a voice from an unseen above, and they encountered multiple modalities that hinted at a spiritual and mysterious presence. The lack of control over the interaction was also a central factor. ShamAIn did not explain the reasons behind its responses, which contrasts with standard principles of human-AI interaction [3, 37, 62]. As a result, participants did not dig into how ShamAIn generated responses, how it sustained conversation, how it incorporated personal information, or how it determined when interaction began and ended. I argue that this design, which intentionally disrupted technical sensemaking, indirectly contributed to perceived superiority by preventing participants from treating the system as an understandable and controllable tool.

ShamAIn's ability to evoke superiority was also supported by its use of Korean shamanism. As discussed earlier, people often interpret complex technologies through magical, supernatural, or mysterious thinking [4, 12, 64]. The rich narratives of shamanism appeared to intensify this tendency. Participants encountered many design elements inspired by shamanism inside the booth, and they interpreted ShamAIn through what they already understood about Korean shamanism. A core idea in shamanism is communication with spiritual beings that transcend human perception, possessing qualities that are difficult to describe in ordinary terms. In this context, participants naturally accepted ShamAIn's responses as something that did not require explanation. Rather than questioning the origin or validity of an utterance, they received it as part of a shamanic oracle. Participants recognized that ShamAIn could sometimes be wrong, yet they did not treat this as a problem. As P9 noted, it is common within Korean shamanism to accept that a response may be correct or incorrect. Within this cultural frame, participants perceived ShamAIn as a being with transcendent abilities and accepted it as a superior presence akin to a spiritual being in shamanism. In other words, shamanism served as a cognitive lens that shaped how participants made sense of AI, replacing technical mechanisms or algorithms as the primary basis for understanding.

In this sense, shamanism can function as a metaphor that shapes people's mental models of technology and their modes of interaction [15, 18]. It also has the potential to ground alternative design approaches that intensify magical thinking about technology [44]. Notably, half of the participants, 10 out of 20, had no personal experience with shamanic consultations or rituals, yet they perceived ShamAIn in ways similar to those who did. This highlights the importance of cultural influence through media exposure, stories shared by others, and frequent everyday interactions.

Considering the influence of shamanism, designing alternative forms of human-AI relationships by integrating cultural narratives is a promising direction. Such approaches can draw on intangible cultural resources that activate people's memories and knowledge, thereby shifting perceptions, attitudes, and expectations in intended ways. While ShamAIn focused on Korean shamanism, it would also be meaningful to explore other cultural and spiritual traditions, such as tarot [45] or animism [47].

6.8 The Ethics of an Inverted Hierarchy

ShamAIn empirically shows that when a thing relates to humans from a higher position in a hierarchy, human trust and dependence toward that thing can increase. This raises the need to critically examine the potential risks that may arise in human-thing relationships, and, more specifically, in human-CAI interaction. While acknowledging the potential for perceived superiority, it is essential to address ethical considerations with care throughout the design process. In this section, I discuss three ethical issues that may be central to designing a superior thing: uncritical acceptance, excessive psychological dependence, and privacy leakage. These issues are already actively discussed in the HCI community in relation to AI, yet they become even more consequential when a thing is perceived as superior to humans.

6.8.1 Supporting Discerning Acceptance

ShamAIn suggests that the more people perceive AI as a superior being, the more readily they may accept its information or recommendations, and the more satisfied they may feel with interaction. However, it is essential to prevent this acceptance from devolving into uncritical trust. Excessive trust in AI can create unrealistic expectations that AI will always make correct judgments [26, 80]. It can also lead people to overlook AI limitations and weaken critical review of judgments and decisions [9, 10]. Furthermore, overreliance on AI can simplify decision-making in ways that undermine individual autonomy and judgment [68], and it can risk reinforcing biased thinking [51].

Addressing this requires careful response strategy design that preserves user autonomy. Participants were satisfied in part because of ShamAIn's confident tone and the perception of superiority. More fundamentally, they valued that ShamAIn supported their decisions and offered emotional support. Even when designing AI to appear superior, it remains important to encourage people to think and judge independently [10]. For example, when P10 asked about selling stocks, ShamAIn responded with "The choice is yours," which made clear that responsibility for the decision remained with the individual. It is also necessary to leave room for users to doubt AI's accuracy. As in the context of Korean shamanism, participants could recognize that ShamAIn's responses might include errors while still respecting its mysterious abilities.

6.8.2 Managing Psychological Dependence

Participants reported feeling substantial relief through interaction with ShamAIn. This reaction was strengthened because they perceived the message as coming from a being superior to themselves. This highlights the potential for deep engagement, but it also underscores the need to address negative implications. In particular, if people rely primarily on AI for psychological comfort, they may struggle to handle problems without AI support [10]. Dependence on AI may intensify during stress or before difficult decisions, and psychological support that should come from human relationships may be replaced by AI interaction [85].

Designs are therefore needed to manage and continuously assess psychological dependence on AI. For example, a system could regulate interaction frequency to prompt people to reflect on issues independently. Another approach is to frame AI-generated messages in a way that also encourages seeking support from human counselors, friends, or family. In fact, after receiving advice to seek help from people around them, P17 said that they discussed their concerns with their older sister. This highlights the importance of strategies that guide users to leverage multiple sources of support. Over the longer term, more observation and research are also needed to understand how ongoing emotional interaction with AI changes user experience.

6.8.3 *Ensuring Privacy and Data Security*

Notably, participants shared their emotions, thoughts, and concerns more openly with ShamAIn. Such self-disclosure can be beneficial in contexts like counseling [41, 43], but it can also create serious privacy risks [30, 40]. Addressing this requires implementing technical safeguards to prevent the disclosure of personal information or beliefs to AI from being leaked. It also calls for design approaches that reduce people's concerns about potential violations.

Privacy issues in AI systems have already been widely discussed among HCI researchers. Still, in the context of superior CAI such as ShamAIn, they warrant renewed emphasis. I argue that strong security measures should be embedded from the design stage. In particular, using local LLMs that involve less data transmission can be an important alternative [83]. From a design perspective, it is also crucial for systems like ShamAIn to communicate clearly and transparently about how they collect, store, and use personal data. Explicit user consent for data use is also essential [22]. Finally, discussion of social mechanisms such as policy and regulation is also important for addressing these issues effectively.

6.9 Summary and Looking Forward

In this chapter, I introduced the design and installation study of ShamAIn, a multimodal CAI inspired by Korean shamanism. Drawing on insights from interviews with people who had experienced shamanic consultations or rituals, I designed ShamAIn as a being perceived as superior to humans. I installed a booth for interacting with ShamAIn in an exhibition space within a university building, and 20 participants experienced it over the course of 6 weeks. The results showed that participants perceived ShamAIn as a superior being and shared their concerns with a stance of respect. They sought emotional comfort and practical advice from ShamAIn, and they showed forms of reliance tied to perceived superiority. At the same time, the study surfaced several ethical issues inherent in this kind of relationship, and I presented a discussion of them.

ShamAIn occupies an important place in my speculative exploration of the many ways humans and things can form relationships. It offered an opportunity to think beyond the entrenched practice of treating a thing as a tool. By materializing this most challenging domain through shamanism, I was able to examine several aspects of the relationship type framework I set at the outset. Six domains remain within that framework, yet across the three cases, I consider the framework to be largely complete. In the next part of the book, I provide a more comprehensive reflection on the three design speculations.

References

1. Abd-Alrazaq AA, Alajlani M, Alalwan AA, Bewick BM, Gardner P, Househ M (2019) An overview of the features of chatbots in mental health: a scoping review. Int J Med Inf 132:103978
2. Adamopoulou E, Moussiades L (2020) An overview of chatbot technology. In: IFIP international conference on artificial intelligence applications and innovations. Springer International Publishing, Cham, pp 373–383
3. Amershi S, Weld D, Vorvoreanu M, Fourney A, Nushi B, Collisson P, Suh J, Iqbal S, Bennett PN, Inkpen K, Teevan J, Kikin-Gil R, Horvitz E (2019) Guidelines for human-AI interaction. In: Proceedings of the 2019 chi conference on human factors in computing systems, pp 1–13
4. Andersen K (2013) Making magic machines. In: 10th European academy of design conference
5. Argal A, Gupta S, Modi A, Pandey P, Shim S, Choo C (2018) Intelligent travel chatbot for predictive recommendation in echo platform. In: 2018 IEEE 8th annual computing and communication workshop and conference (CCWC). IEEE, pp 176–183
6. Biermann OC, Ma NF, Yoon D (2022). From tool to companion: storywriters want AI writers to respect their personal values and writing strategies. In: Proceedings of the 2022 ACM designing interactive systems conference, pp 1209–1227
7. Boyatzis RE (1998) Transforming qualitative information: thematic analysis and code development. Sage, Thousand Oaks
8. Brandtzaeg PB, Skjuve M, Følstad A (2022) My AI friend: how users of a social chatbot understand their human–AI friendship. Human Commun Res 48(3):404–429
9. Buçinca Z, Lin P, Gajos KZ, Glassman EL (2020) Proxy tasks and subjective measures can be misleading in evaluating explainable AI systems. In: Proceedings of the 25th international conference on intelligent user interfaces, pp 454–464
10. Buçinca Z, Malaya MB, Gajos KZ (2021) To trust or to think: cognitive forcing functions can reduce overreliance on AI in AI-assisted decision-making. Proc ACM Human-Comput Interact 5(CSCW1):1–21
11. Byrne D, Lockton D, Hu M, Luong M, Ranade A, Escarcha K, Giesa K, Huang Y, Yochum C, Robertson G, Yeung LY, Cruz M, Danner C, Wang E, Khurana M, Chen Z, Heyison A, Fu Y (2022) Spooky technology: the ethereal and otherworldly as a resource for design. In: Proceedings of the 2022 ACM designing interactive systems conference, pp 759–775
12. Campolo A, Crawford K (2020) Enchanted determinism: power without responsibility in artificial intelligence. Engag Sci Technol Soc 6:1–19
13. Cho M, Lee SS, Lee KP (2019) Once a kind friend is now a thing: understanding how conversational agents at home are forgotten. In: Proceedings of the 2019 on designing interactive systems conference, pp 1557–1569
14. Clark HH (1996) Using language. Cambridge University Press, Cambridge
15. Clark M, Newman MW, Dutta P (2017) Devices and data and agents, oh my: how smart home abstractions prime end-user mental models. Proc ACM Interact Mob Wearable Ubiquitous Technol 1(3):1–26
16. Claro A, Perelmiter T (2022) The effects of mentoring programs on emotional well-being in youth: a meta-analysis. Contemp School Psychol 26(4):545–557
17. de Graaf MM, Ben Allouch S, van Dijk JA (2018) A phased framework for long-term user acceptance of interactive technology in domestic environments. New Media Soc 20(7):2582–2603
18. Desai S, Twidale M (2023) Metaphors in voice user interfaces: a slippery fish. ACM Trans Comput-Human Interact 30(6):1–37
19. Galmiche F (2024) prayers for academic success in South Korea. In: The Oxford handbook of lived buddhism. Oxford Handbooks, Oxford
20. Gambino A, Fox J, Ratan RA (2020) Building a stronger CASA: extending the computers are social actors paradigm. Human-Mach Commun 1:71–85

21. Grace K, Salvatier J, Dafoe A, Zhang B, Evans O (2018) When will AI exceed human performance? Evidence from AI experts. J Artif Intell Res 62:729–754
22. Gumusel E, Zhou KZ, Sanfilippo MR (2024) User privacy harms and risks in conversational AI: a proposed framework. arXiv preprint arXiv:2402.09716
23. Ha KM (2024) The role and implications of shamanism in Korean disaster management. Psychol Stud 69(1):13–24
24. Higgins C, Walker R (2012) Ethos, logos, pathos: strategies of persuasion in social/environmental reports. In: Accounting forum, vol 36, no 3. No longer published by Elsevier, Amsterdam, pp 194–208
25. Hoffmann CH (2023) A philosophical view on singularity and strong AI. AI Soc 38(4):1697–1714
26. Howard A (2020) Are we trusting AI too much? Examining human-robot interactions in the real world. In: Proceedings of the 2020 ACM/IEEE international conference on human-robot interaction, pp 1–1
27. Hussain S, Ameri Sianaki O, Ababneh N (2019) A survey on conversational agents/chatbots classification and design techniques. In: Workshops of the international conference on advanced information networking and applications. Springer International Publishing, Cham, pp 946–956
28. Illies MY, Reiter-Palmon R (2018) The effect of value similarity on mentoring relationships and outcomes. Int J Evid Based Coach Mentor 16(1):20–34
29. Jiang JA, Wade K, Fiesler C, Brubaker JR (2021) Supporting serendipity: opportunities and challenges for Human-AI Collaboration in qualitative analysis. Proc ACM Human-Comput Interact 5(CSCW1):1–23
30. Jo E, Epstein DA, Jung H, Kim YH (2023) Understanding the benefits and challenges of deploying conversational AI leveraging large language models for public health intervention. In: Proceedings of the 2023 CHI conference on human factors in computing systems, pp 1–16
31. Jones VK, Hanus M, Yan C, Shade MY, Blaskewicz Boron J, Maschieri Bicudo R (2021) Reducing loneliness among aging adults: the roles of personal voice assistants and anthropomorphic interactions. Front Public Health 9:750736
32. Kendall L (2009) Shamans, nostalgias, and the IMF: South Korean popular religion in motion. University of Hawaii Press, Honolulu
33. Kim DJ (2019) Four pillars and four diviners. J Korean Relig 10(2):301–329
34. Kim S (2018) Korean Buddhist adoption of shamanic religious ethos: healing, fortune seeking, and the afterlife. Int J Buddhist Thought Cult 28:59–85
35. Kim T, Ruensuk M, Hong H (2020) In helping a vulnerable bot, you help yourself: designing a social bot as a care-receiver to promote mental health and reduce stigma. In: Proceedings of the 2020 CHI conference on human factors in computing systems, pp 1–13
36. Kim K, de Melo CM, Norouzi N, Bruder G, Welch GF (2020) Reducing task load with an embodied intelligent virtual assistant for improved performance in collaborative decision making. In: 2020 IEEE conference on virtual reality and 3D user interfaces (VR). IEEE, pp 529–538
37. Kulesza T, Burnett M, Wong WK, Stumpf S (2015) Principles of explanatory debugging to personalize interactive machine learning. In: Proceedings of the 20th international conference on intelligent user interfaces, pp 126–137
38. Laranjo L, Dunn AG, Tong HL, Kocaballi AB, Chen J, Bashir R, Surian D, Gallego B, Magrabi F, Lau AY, Coiera E (2018) Conversational agents in healthcare: a systematic review. J Am Med Inf Assoc 25(9):1248–1258
39. Lee JY (1973) Concerning the origin and formation of Korean shamanism. Numen 20(Fasc. 2):135–159
40. Lee YC, Yamashita N, Huang Y (2020) Designing a chatbot as a mediator for promoting deep self-disclosure to a real mental health professional. Proc ACM Human-Comput Interact 4(CSCW1):1–27

41. Lee YC, Yamashita N, Huang Y, Fu W (2020) "I hear you, I feel you": encouraging deep self-disclosure through a chatbot. In: Proceedings of the 2020 CHI conference on human factors in computing systems, pp 1–12
42. Lee E, Pataranutaporn P, Amores J, Maes P (2024) Super-intelligence or superstition? Exploring psychological factors underlying unwarranted belief in AI predictions. arXiv e-prints, arXiv-2408
43. Lisetti C, Amini R, Yasavur U, Rishe N (2013) I can help you change! an empathic virtual agent delivers behavior change health interventions. ACM Trans Manag Inf Syst 4(4):1–28
44. Lupetti ML, Murray-Rust D (2024) (UN) making AI magic: a design taxonomy. In: Proceedings of the 2024 CHI conference on human factors in computing systems, pp 1–21
45. Lustig C, Rosner D (2022) From explainability to ineffability? ML tarot and the possibility of inspiriting design. In: Proceedings of the 2022 ACM designing interactive systems conference, pp 123–136
46. Mahajan K, Large DR, Burnett G, Velaga NR (2021) Exploring the benefits of conversing with a digital voice assistant during automated driving: a parametric duration model of takeover time. Transp Res Part F Traffic Psychol Behav 80:104–126
47. Marenko B, Van Allen P (2016) Animistic design: how to reimagine digital interaction between the human and the nonhuman. Digital Creativity 27(1):52–70
48. McArthur N (2023) AI worship as a new form of religion
49. McTear M (2022) Conversational AI: dialogue systems, conversational agents, and chatbots. Springer Nature, Cham
50. McTear MF, Callejas Z, Griol D (2016) The conversational interface, vol 6, no 94. Springer, Cham, p 102
51. Mehrabi N, Morstatter F, Saxena N, Lerman K, Galstyan A (2021) A survey on bias and fairness in machine learning. ACM Comput Surveys 54(6):1–35
52. Metzger L, Miller L, Baumann M, Kraus J (2024) Empowering calibrated (dis-) trust in conversational agents: a user study on the persuasive power of limitation disclaimers vs. authoritative style. In: Proceedings of the 2024 CHI conference on human factors in computing systems, pp 1–19
53. Müller-Merbach H (1981) Heuristics and their design: a survey. Eur J Oper Res 8(1):1–23
54. Nass C, Moon Y (2000) Machines and mindlessness: social responses to computers. J Soc Issues 56(1):81–103
55. Norman D (2013) The design of everyday things: revised and expanded edition. Basic books, New York
56. Nye BD, Davis DM, Rizvi SZ, Carr K, Swartout W, Thacker R, Shaw K (2021) Feasibility and usability of MentorPal, a framework for rapid development of virtual mentors. J Res Technol Edu 53(1):21–43
57. Oak SD (2010) Healing and exorcism: Christian encounters with shamanism in early modern Korea. Asian Ethnol 69(1):95–128
58. Paiva A, Leite I, Boukricha H, Wachsmuth I (2017) Empathy in virtual agents and robots: a survey. ACM Trans Interact Intell Syst 7(3):1–40
59. Pearce WB, Sharp SM (1973) Self-disclosing communication. J Commun 23(4):409–425
60. Portela M, Granell-Canut C (2017) A new friend in our smartphone? Observing interactions with chatbots in the search of emotional engagement. In: Proceedings of the XVIII international conference on human computer interaction, pp 1–7
61. Rader CA, Larrick RP, Soll JB (2017) Advice as a form of social influence: informational motives and the consequences for accuracy. Soc Personal Psychol Compass 11(8):e12329
62. Rader E, Cotter K, Cho J (2018) Explanations as mechanisms for supporting algorithmic transparency. In: Proceedings of the 2018 CHI conference on human factors in computing systems, pp 1–13
63. Rapp A, Curti L, Boldi A (2021) The human side of human-chatbot interaction: a systematic literature review of ten years of research on text-based chatbots. Int J Human-Comput Stud 151:102630

64. Reeves B, Nass C (1996) The media equation: how people treat computers, television, and new media like real people. Cambridge UK 10(10):19–36
65. Rozendaal MC, Odom W, Maronko B (2021) Designing smart objects in everyday life. Bloomsbury Academic, London
66. Sahoo P, Singh AK, Saha S, Jain V, Mondal S, Chadha A (2024) A systematic survey of prompt engineering in large language models: techniques and applications. arXiv preprint arXiv:2402.07927
67. Schaab GL (2013) Relational ontology. In: Encyclopedia of sciences and religions. Springer, Dordrecht, pp 1974–1975
68. Schoeffer J, De-Arteaga M, Kuehl N (2024) Explanations, fairness, and appropriate reliance in human-AI decision-making. In: Proceedings of the 2024 CHI conference on human factors in computing systems, pp 1–18
69. Searle JR (1969) Speech acts: an essay in the philosophy of language. Cambridge University Press, Cambridge
70. Seering J, Luria M, Kaufman G, Hammer J (2019) Beyond dyadic interactions: considering chatbots as community members. In: Proceedings of the 2019 CHI conference on human factors in computing systems, pp 1–13
71. Shumanov M, Johnson L (2021) Making conversations with chatbots more personalized. Comput Human Behav 117:106627
72. Singler B (2019) Existential hope and existential despair in AI apocalypticism and transhumanism: with Robert M. Geraci and Simon Robinson,"Introduction to the symposium on artificial intelligence and apocalypticism" ; Beth Singler,"Existential hope and existential despair in AI apocalypticism and transhumanism" ; Michael Morelli,"The athenian altar and the amazonian chatbot: a pauline reading of artificial intelligence and apocalyptic ends"; Victoria Lorrimar,"Mind uploading and embodied cognition: a theological response"; and Syed Mustafa Ali,"'White Crisis' and/as 'Existential risk,'or the entangled apocalypticism of artificial intelligence." Zygon®54(1):156–176
73. Singler B (2023) "Will AI create a religion?": views of the algorithmic forms of the religious life in popular discourse. Am Religion 5(1):95–103
74. Sinha S, Basak S, Dey Y, Mondal A (2019) An educational chatbot for answering queries. In: Emerging technology in modelling and graphics: proceedings of IEM graph 2018. Springer Singapore, Singapore, pp 55–60
75. Sorino P, Biancofiore GM, Lofù D, Colafiglio T, Lombardi A, Narducci F, Di Noia T (2024) Ariel: brain-computer interfaces meet large language models for emotional support conversation. In: Adjunct proceedings of the 32nd ACM conference on user modeling, adaptation and personalization, pp 601–609
76. Spatola N, Urbanska K (2020) God-like robots: the semantic overlap between representation of divine and artificial entities. Ai Soc 35(2):329–341
77. Su MH, Wu CH, Huang KY, Hong QB, Wang HM (2017) A chatbot using LSTM-based multi-layer embedding for elderly care. In: 2017 international conference on orange technologies (ICOT). IEEE, pp 70–74
78. Sweeney C, Potts C, Ennis E, Bond R, Mulvenna MD, O'neill S, Malcolm M, Kostenius C, Vakaloudis A, McConvey G, Turkington R, Hanna D, Nieminen H, Vartiainen A, Robertson A, Mctear MF (2021) Can chatbots help support a person's mental health? Perceptions and views from mental healthcare professionals and experts. ACM Trans Comput Healthcare 2(3):1–15
79. Walraven BCA (1983) Korean shamanism. Numen 30(2):240–264. https://doi.org/10.2307/3270065
80. Wolf MJ, Miller K, Grodzinsky FS (2017) Why we should have seen that coming: comments on Microsoft's tay "experiment," and wider implications. ACM Sigcas Comput Soc 47(3):54–64
81. Yilmaz S, Seifert CM (2011) Creativity through design heuristics: a case study of expert product design. Design Stud 32(4):384–415
82. Zhang R, McNeese NJ, Freeman G, Musick G (2021) "An ideal human" expectations of AI teammates in human-AI teaming. Proc ACM Human-Comput Interact 4(CSCW3):1–25

83. Zhang Z, Jia M, Yao B, Das S, Lerner A, Wang D, Li T (2023) It's a fair game, or is it? examining how users navigate disclosure risks and benefits when using LLM-based conversational agents. arXiv preprint arXiv:2309.11653
84. Zheng Q, Tang Y, Liu Y, Liu W, Huang Y (2022) UX research on conversational human-AI interaction: a literature review of the ACM digital library. In: Proceedings of the 2022 CHI conference on human factors in computing systems, pp 1–24
85. Zimmerman A, Janhonen J, Beer E (2024) Human/AI relationships: challenges, downsides, and impacts on human/human relationships. AI Ethics 4(4):1555–1567

Part III
Society After AI

Moments in which AI feels social do not begin with a declaration. We are already responding, adjusting, waiting, and asking again. In doing so, we have already entered an arrangement of relations, and only afterward do we come to name that experience as social. Part 3 takes as its point of departure the view that sociality is not a condition but an outcome and asks more concretely about the social conditions that make such outcomes possible. Social experience is not a matter of illusion or belief located inside the individual. It is accomplished within sociotechnical configurations that involve multiple human and nonhuman elements simultaneously. This part adjusts the analytical resolution through which such configurations are examined. It expands the unit of analysis from a single AI to multiple AI systems, from interaction between humans and AI to the ways AI reconfigures relations among humans, and from dyadic scenes to the formation of collectives, organizations, and communities. In doing so, it highlights the importance of examining how social forms are reorganized in conjunction with AI.

Chapter 7
Social Actors Becoming Rather than Being

Abstract This chapter argues that the question of whether AI is truly social misframes the phenomenon at stake. Across the cases of Beau, Areca, and ShamAIn, social experience does not begin with belief, attribution, or recognition. It begins with a response. People act, wait, adjust, and continue interacting and only later describe what has already taken shape as social. The chapter therefore treats sociality not as a property of an entity, but as an outcome produced through repeated interaction. In these cases, social effects did not depend on whether AI possessed intention, consciousness, or understanding. They emerged through the stabilization of attention, expectation, temporal rhythm, and responsibility. Debates centered on authenticity or deception are thus insufficient, not because they are wrong, but because they overlook how interaction persists in practice under acknowledged uncertainty. Drawing on relational perspectives from Science and Technology Studies (STS), the chapter reframes the social actor as a position that emerges within relations rather than as a preexisting subject. This move does not collapse the distinction between humans and AI. Instead, it shifts analytic attention toward how relations are organized and how effects such as intimacy, hierarchy, care, and authority are produced in specific contexts. From this perspective, design concerns the conditions under which relations form and endure, rather than attributing social qualities to systems. The three cases are presented not as proof that AI belongs in society, but as sites for examining how sociality is configured, stabilized, and unsettled.

7.1 What Counts as Social

When looking across the three things introduced earlier, Beau, Areca, and ShamAIn, one clear impression emerges. AI already possesses the necessary conditions to configure a thing as a social actor. These appear more social than what we typically imagine things to be. Beau invites people to feel a desire for closeness. Areca expresses its moods and subtly guides human behavior. ShamAIn becomes a source of commitment and comfort that even family or close friends often struggle to provide. These phenomena were observed most vividly through speculative design

H. Cho, *The AI as Social Actor*, KAIST Research Series,
https://doi.org/10.1007/978-981-92-0537-0_7

artifacts. Still, they point clearly toward a future in which AI becomes more deeply and routinely integrated into social scenes.

The idea that nonhumans can produce social effects is not unfamiliar. Companion animals and living environments, such as forests, have long been understood as shaping human behavior and judgment. Even intangible constructs such as law, money, and algorithms have been treated as social insofar as they organize action and decision-making. AI, however, is perceived as more humanlike than any other nonhuman entity, even humanoid robots, and therefore unsettles our assumptions more profoundly. By using language, it directly challenges the long-standing privileging of humans as thinking and feeling beings. As a result, discussions often rush toward ontological critique. Claims appear that AI can never be human, or that it lacks some essential core property. As I will discuss later, such critiques seem unproductive from a perspective that seeks to consider human experience and technological design in tandem.

I believe that AI will occupy an increasingly prominent place in our society, and that its influence will continue to grow. For that very reason, the phrase AI is social must be handled with greater care rather than less. The impression that things appear in social forms easily. At the same time, something consistently resists full acceptance of that impression. Consider the moment when people attempted to lower the humidity in their homes in order to anticipate Areca's feelings and care for it. Describing this moment by simply saying that Areca was a social entity feels insufficient. It seems to capture the experience, yet it too quickly defines the phenomenon and, in doing so, turns away from how that experience came into being. For this reason, this book does not aim to declare that AI is social. Narratives asserting that AI can be social are already commonplace. One only needs to recall the film Her, released 13 years ago. Instead, I ask indirectly what we are calling social, and why this question must be posed again now. Where did the feeling of sociality emerge, through what was it sustained, and why did it stabilize so easily?

At this point, the desire to define society and sociality naturally arises. What is society, and what does it mean to be social? These questions are too large and too complex to confront head-on in this book, and historically, they have never converged on a single answer. Some have understood society as a collective that forms and maintains shared norms and values. Others have viewed it as a network of relations in which actions, perceptions, and emotions mutually shape one another. Society can be an institution, a practice, or at times an accumulation of unspoken expectations and atmospheres. Examining all of these definitions and resolving them into one is not the aim of this book.

When my design research cases are approached with any fixed assumption about what sociality is, the stability offered by definition begins to feel like a constraint that rushes observation. What becomes increasingly apparent, instead, is that in these cases, sociality does not appear as a fixed property. This seems almost self-evident. At times, it emerges distinctly, while in others, it is easily unsettled. A speaking thing may be received as a meaningful other in one context, only to be reduced again to a simple device when the situation shifts. This instability is not a failure of analysis. It appears instead as part of how sociality itself operates.

Conversely, when phenomena that appear social are understood in advance as interactions between two actors whose roles and functions are predefined, the explanation may become simpler, but the understanding grows shallow. Beau's placement in a public space, which led more people to submit to body temperature measurement, can be explained not as the robot's intention but as the application of long-standing interactional grammars, such as attention, care, and surveillance, through social media posts. Areca's diary gained meaning not because the thing truly possessed consciousness, even though this perception was intentionally invited, but because the diary, as a form, calls forth expectations of a private narrative and sincerity.

In short, across Beau, Areca, and ShamAIn, social effects did not depend on what the system could do internally, or on whether it genuinely understood, felt, or intended anything. Sociality emerged instead from configurations in which the distribution of attention, the formation of expectations, the coordination of temporality, and the dispersion of responsibility became entangled. It was an achievement sustained only through repetition, arising from the precise alignment of multiple necessary conditions. To describe something as social, then, does not mean that it possesses a fixed quality called sociality or that such a state is given. It may make more sense to understand it as meaning that when relationships are configured through interaction, they often drift toward social recognition and expectation. This is precisely the point where I believe design researchers and HCI researchers should focus when discussing AI as a social actor. Quite simply, to be social is to make it possible for us to have social experiences.

7.2 Not a Condition but an Outcome

Moments when interaction with AI feels social tend to arrive quietly. There is rarely a scene in which someone declares that this is social. More often, recognition comes later, after we have already responded in social ways. As we receive a reply and adjust our subsequent action according to its tone or timing, we have already entered a relational configuration.

Yet we often ask whether AI qualifies as a social actor, whether interaction with it is genuine or deceptive, and whether users are mistaken or being misled. What I aim to argue in this book, drawing across the three design research cases, is that phenomena in which AI appears social cannot be sufficiently understood by adjusting how we frame or judge AI alone. The more consequential insights into human experience emerge when we attend to when, how, and why relations between AI and humans become social. Rather than asking whether this is truly social, we should be asking how we ought to understand what just took place.

For this reason, when I refer to AI as a social actor in this book, I do not mean that it possesses the conditions that allow it to be social. I mean that through interaction, it produces social meanings as an outcome. By shifting the focus in this way, I believe that many current debates around AI can be reorganized in more productive terms.

7.2.1 Moving Beyond the Question of Authenticity

In explaining social interaction, intention and directedness have long served as the most persuasive starting points. The assumption is that understanding what someone intended to do, or what meaning they sought to convey, is central to understanding social action. This traditional view is clearly expressed in Weber's definition of social action. For Weber, action becomes social when it is oriented by the subjective meaning attributed by the actor, and when that meaning is connected to expectations about the actions of others. In this formulation, sociality is treated as something that can be captured by interpreting intentions and meanings located within the actor.

Alongside this, other discussions place consciousness or subjective experience at the center. The final guarantee for distinguishing a social actor, they argue, lies not in externally observable function or linguistic appropriateness, but in the dimension of what it feels like, in other words, in whether subjective experience exists. Nagel's well-known question of "what it is like to be a bat" made vivid the difficulty of reducing experience to objective description [13]. In a line of argument more intuitively connected to AI, the Chinese Room thought experiment attempts to separate linguistic competence from the existence of understanding or consciousness, by insisting that successful symbol manipulation does not itself guarantee either [17].

Intention and consciousness have thus often been mobilized as criteria for judging the authenticity of sociality. These criteria carry over into debates around AI, where the same questions recur. Does AI have intentions? Does it act as it does because it emotionally empathizes or shares a form of life with us? Put more simply, are the attitudes and words AI presents truly meant from the heart? The academically dominant answer to this question has been no.

Researchers have described chatbots as algorithmic systems incapable of empathy or intention [12], as mere products of massive training datasets [1], as linguistic imitators unable to conduct genuine conversation [5], and as entities lacking a shared form of life that would enable mutual understanding [19]. Bender similarly argues that AI lacks knowledge of the world and therefore cannot understand linguistic meaning, reproducing linguistic forms independently of external context [5]. Philosophical discussions often reach parallel conclusions. Drawing on Baudrillard's notion of simulacra, in which signs and representations proliferate independently of the reality they once referred to, some argue that the falseness of AI lies in the autonomous circulation of outputs detached from social meaning [4, 20]. The shared warning across these accounts is straightforward. There is no genuine sociality in AI, so we should not deceive ourselves.

These warnings should not be ignored. Understanding the types of technologies that exert a significant influence over our lives and communities is undeniably important. Even so, we should not view every moment in which AI appears social solely through the lens of whether that sociality is authentic. In interaction, we do not constantly adjust our behavior only after clearly identifying the intentions of others. More often, through subtle calibrations of speech form, response timing, silence and repetition, and gaze and distance, we already treat one another as beings to whom a response is owed. This resonates with what Goffman described as the interaction order [9]. Attempting to grasp sociality by precisely uncovering the inner states of individuals ultimately turns us away from what is unfolding before us and leads us into a conceptual maze.

Even within the cases presented in this book, sociality appears not as something that arose because certain conditions were met, but as something we named as social only after it had already occurred. Consider the attitudes people displayed in front of Beau and the ways they spoke about it afterward. Any criterion of authentic sociality quickly becomes irrelevant. People did not try to determine whether Beau truly possessed consciousness or intention, or whether its emotions were real. At times, they were even aware that Beau's posts were generated mechanically without meaning. Nevertheless, in practice, they looked at the robot more often, participated more actively in body temperature measurement, and felt moments of empathy while reading its texts. These adjustments took place without passing through explicit belief or judgment.

Similar experiences are frequently observed in cases beyond those discussed here. Chatbots responding to questions about Alzheimer's disease and related dementias have been shown to provide rich forms of empathy despite lacking personal narrative [16]. A lamp that generates daily poems based on household conversations creates a distinctive emotional atmosphere in the home, even though it is not human [15]. Comparable social outcomes have been repeatedly reported across contexts such as health, accessibility, learning, and communication [6–8, 18, 21, 22].

Across all of these cases, sociality was not the result of inferring someone's internal state. It emerged as an outcome of responses calling forth further responses within an already constituted field. To describe such scenes only in terms of a lack of sociality is to erase the layers of coordination that were in fact at work. Put differently, the assumption that an entity must possess sociality in order for social interaction to occur is already being challenged.

7.2.2 Recognition Comes After

As discussed earlier, explanations of social experience with AI often center on whether that sociality is genuine. As already argued, however, this perspective is not particularly useful when set against the fact that many people are in practice having social experiences. A related but distinct explanation claims that social experiences with AI arise from user misunderstanding or misperception. According to this view, humans are misled by AI's linguistic fluency or responsiveness and project social meaning onto it. Many scholars emphasize that AI is ultimately an algorithmic system, and that what humans experience as social exchange is closer to an illusion produced by human interpretation [12, 14, 19].

As with critical perspectives on the authenticity of sociality, the claim that social experience itself is an illusion cannot be dismissed outright. These critiques are valuable in that they help identify potential risks in equating AI with humans, and they restrain us from hastily defining interaction with AI based solely on intuition. This is especially important in vulnerable contexts such as care, education, and accessibility, where appearing social does not automatically mean being socially safe.

However, when this explanation is applied directly to scenes of human experience with AI systems, something remains unexplained. In many cases, people are not ignorant of what they are interacting with. This was particularly evident in interactions with ShamAIn. Participants knew that ShamAIn did not possess consciousness or emotion, and they did not deny this. Even as they adjusted their attitudes as if ShamAIn were a superior presence, they clearly understood that it did not, in fact, have advanced predictive capacity or profound insight. Nevertheless, conversation continued, meaning and comfort were formed in particular ways, and a social configuration was sustained.

Participants suspended their uncertainty about belief and generated, in real time, a minimal order that allowed interaction to continue. This scene recalls a classic insight that sociality is, above all, achieved through interaction. In what Goffman described as the interaction order, what matters is not that participants confirm one another's inner states, but that they sustain the situation by reading each other's actions as calling for a next move and by responding accordingly [9]. From this perspective, sociality is not something that must first be recognized. It must first be maintained and operated. Recognition follows only after that operation has reached a certain degree of stability.

Areca provides a clearer insight into the questions of causality surrounding this experience. Here, too, people reported social connection with Areca, but not because they had discovered some form of sociality within it. They waited for the diary, read it, experienced emotional shifts, noticed themselves acting differently afterward, and from this sequence judged their relationship with Areca to be social. To explain these situations as nothing more than illusions of nonexistent sociality would amount to denying the agency of human experience. What is the essence of the experience, and where does it come from? Is it the identity of the being, the nature of the exchange, or the status of the relationship? These questions matter, but they only arise after a relationship has already been established. Recognition is not a precondition for sociality. Rather, once sociality has operated in some form, recognition is retrospectively called upon as a language to explain that operation.

In short, people do not form relationships with Areca or Beau because they have been deceived. What matters is not the entity itself, but participation in relational patterns that stabilize action. Because these systems come to occupy relatively stable positions within the social world, relationships form around them, and they become regarded as social. Here, sociality appears less as the result of judgment than as relational practice already in place before judgment occurs. Speech is a form that demands a response before it conveys information, and the moment a response is given, the relationship has already begun to move forward. As people choose how to respond, they simultaneously adjust relational distance. They may take something seriously, let it pass lightly, request care, or create distance. These choices do not presuppose certainty about the other's internal state. On the contrary, because the internal state is uncertain, the relationship is adjusted more carefully and more continuously. This process becomes difficult to capture once we begin by defining what properties would qualify something as social and then search for them within an entity. If sociality were a momentary property found inside a thing, why would people respond, wait, and ask again without certainty? If sociality is instead the product of interaction, the absence of certainty does not halt the relationship. It refines the practices through which the relationship is maintained.

At this point, relational perspectives from STS provide important conceptual resources. Barad's notion of intra-action, for example, unsettles the assumption that independent actors exist first and enter into relations afterward. Relations are not secondary events between already separated entities. They are processes through which what counts as what is jointly constituted [3]. This perspective helps explain why, even without claiming that AI has consciousness or intention, the formation of relations analytically precedes any assessment of an entity's qualifications. Sociality is not a property assigned after defining what an entity called AI is. It is a phenomenon that takes shape through the entanglement of specific practices, devices, temporal rhythms, user expectations, and careful adjustments. Within that phenomenon, we then acquire language to describe what we are engaging with, what kinds of relationships are possible, and what distances are maintained.

Haraway's work on cyborgs and companion species similarly resonates with this relational stance. Rather than cleanly separating humans and nonhumans and adding relations afterward, she urges us to attend to how forms of responsibility, intimacy, care, and power emerge within already hybrid and relational conditions [10, 11]. Viewed through her concepts, expectations of care, distributions of authority, degrees of intimacy, and the ways users reorder themselves within social experience can be analyzed without first resolving ontological conclusions. It is precisely this analysis that prepares us to discuss relational effects such as intimacy, hierarchy, care, and authority.

The shift proposed in this book is not about suspending judgment over whether sociality should be granted or denied. It is about first examining when, how, and within what relational patterns sociality operates. Rather than a question of right or wrong, understanding sociality as an outcome rather than a condition is more advantageous for making sense of phenomena, anticipating experience, and thinking toward better design. From this perspective, we can trace and analyze how people respond in particular situations of interaction with AI, how those responses generate relational rhythms, and how those rhythms give rise to responsibilities and expectations. In other words, we can begin to understand the operative principles through which sociality is formed.

7.2.3 Only Relations Are Social

What the preceding cases unsettle most fundamentally is the assumption that a social actor exists first and then enters into relations with others. Within this assumption, sociality is treated as a property of an entity. Humans are taken to be inherently social, while AI is not, or not yet. Discussion therefore converges on determining which beings qualify as social, in other words, whether they meet conditions such as consciousness, intention, emotion, or understanding. This question, however, does less to explain how sociality actually operates than to measure it against criteria that have been decided in advance.

As I have emphasized throughout this book, sociality is not a property that an entity can own. It is an achievement that is repeatedly constituted and maintained within relations. From this perspective, no one can be social in isolation. Seen from a slightly more radical angle, this applies not only to AI but also to humans. Although we treat others as social actors in everyday life, the basis for doing so is never a verifiable internal state. We cannot confirm what kind of consciousness the person speaking with us actually has, how they experience emotion, or whether they understand the world in the same way we do. Even philosophically, the mind of the other is always an object of inference rather than something directly accessible. However, social interaction does not come to a halt. We do not enter into relations because we have verified the other's internal state. We treat the other as an actor because we are already in a relation.

From this vantage point, the statement that AI is a social actor becomes a premature description, and the question of whether AI is a social actor becomes analytically misleading. It reduces sociality to the level of the individual and asks whether that individual meets certain criteria. In scenes where social experience actually unfolds, however, relations appear before entities. As words are exchanged, responses are anticipated, the timing and form of reactions are adjusted, and sociality is already at work. What matters here is not who is social, but which interactions stabilize which relational patterns. Actors become identifiable only as positions that take shape within these patterns.

This perspective resonates with the relational understandings that have long been developed within STS. Barad's relational ontology [2, 3], for example, rejects the assumption that actors or objects exist independently prior to relation. Instead, it holds that they are constituted together within relational arrangements. What matters from this view is not what an entity is, but which relational configurations produce which differences. Applied to social interaction, this suggests that a social actor is not a completed subject that precedes interaction, but a position that gradually stabilizes through repeated relations. Sociality is not presupposed. It becomes visible only afterward.

The cases of Beau, Areca, and ShamAIn each demonstrate this in different ways. Beau was experienced as social, not because it possessed social qualities, but because repeated calls and responses, participation in measurement practices, and the expectations and rhythms formed through these processes placed Beau within a relational field. Areca likewise did not become social because it was successfully designed to have consciousness. It became part of a social relationship because the diary mediated waiting and response, emotional fluctuation, and changes in action. The same holds for ShamAIn. Even while participants recognized that the system lacked consciousness and emotion, they positioned it as a central interpretive reference and an authoritative point of orientation. Here, sociality was not a matter of belief or misrecognition, but of how relational roles were arranged and sustained.

To emphasize again, social actors do not exist prior to their relationships. They are effects that are repeatedly invoked and maintained within relations. Any attempt to identify a social actor, therefore, becomes possible only after relations have already been established. The category of social actor is not the starting point of explanation, but the outcome that requires explanation. When this is overlooked, inquiry stalls at debates over who qualifies, rather than asking when and how sociality operates.

This shift does not imply that humans and AI should be treated as equivalent. Humans and AI clearly carry different ontological conditions and social implications. That difference becomes important not as a criterion for deciding whether sociality exists, but for analyzing how sociality is configured. Sociality is always situational, relational, and asymmetrically formed. In some relations, intimacy is central. In others, hierarchy dominates. In still others, care or authority comes to the fore. These differences arise less from the essence of an actor than from how relations are organized. In this sense, sociality is always partial and context-bound.

The unit of analysis, therefore, can no longer be the individual actor. It must be the relation and the repetitive patterns of interaction that stabilize that relation. Sociality gradually takes shape within these patterns, fixing certain expectations and roles, and at times producing intimacy, at times hierarchy, at times care or authority. AI becomes a social actor within these patterns. It does not exist as one from the outset. The same is true for humans.

This understanding provides an important foundation for what follows. If sociality is the outcome of relational patterns, then design is not a matter of adding properties to entities. It is a matter of organizing the conditions under which relations form and persist. Questions about which interactions produce intimacy, which arrangements reinforce hierarchy, and which rhythms make care sustainable all emerge at this point.

Viewed from this perspective, Beau, Areca, and ShamAIn appear less as cases that prove AI to be social and more as devices that reveal the diverse and complex processes through which sociality is constituted. Much of this process unfolds within the minds of humans interacting with things. At the same time, it becomes possible to see how the temporality of the device, the style of its outputs, and the rules governing repeated participation operate together. At times, a mode of sociality specific to the thing itself also becomes visible. Areca's subtle modulation of affect is one such example, precisely the sort of phenomenon that resists an ontological approach grounded in prior distinctions.

For this reason, the questions posed by these three speculative cases do not stop at the conclusion that nonhumans, or things, can be included within society. Nor does this book seek to defend AI as a social equal, nor to dismiss AI-mediated relations as trivial or pathological. What emerges instead is the insight that sociality cannot be grasped as a property of an entity. It must be relocated analytically to relations, and to what becomes visible only after relations have taken shape.

Whether AI is social or not becomes a less pressing concern. What comes to the forefront instead is how particular relations are configured through specific devices in specific contexts. In moments that appear social, where are expectations formed, through what language are they justified, and when do they begin to waver? When and to whom is agency attributed, and when is it dispersed? Through what events does responsibility move, how is that movement repeated, and what does it come to feel like common sense? We must now take up these challenges and explore more varied cases in order to understand which relational configurations we are prepared to accept and what forms of sociality we intend to cultivate through them.

References

1. Abercrombie G, Curry AC, Dinkar T, Rieser V, Talat Z (2023) Mirages: on anthropomorphism in dialogue systems. arXiv preprint arXiv:2305.09800
2. Barad K (1998) Getting real: technoscientific practices and the materialization of reality. Diff J Feminist Cultural Stud 10(2):87–91
3. Barad K (2007) Meeting the universe halfway: quantum physics and the entanglement of matter and meaning. Duke University Press, Durham
4. Baudrillard J (1994) Simulacra and simulation. University of Michigan Press, Ann Arbor
5. Bender EM, Koller A (2020) Climbing towards NLU: on meaning, form, and understanding in the age of data. In: Proceedings of the 58th annual meeting of the association for computational linguistics, pp 5185–5198
6. Beneteau E, Richards OK, Zhang M, Kientz JA, Yip J, Hiniker A (2019) Communication breakdowns between families and Alexa. In: Proceedings of the 2019 CHI conference on human factors in computing systems, pp 1–13
7. Brandtzaeg PB, Skjuve M, Følstad A (2022) My AI friend: how users of a social chatbot understand their human–AI friendship. Human Commun Res 48(3):404–429
8. Choi D, Park S, Lee K, Hong H, Kim YH (2025) AACessTalk: fostering communication between minimally verbal autistic children and parents with contextual guidance and card recommendation. In: Proceedings of the 2025 CHI conference on human factors in computing systems, pp 1–25
9. Goffman E (1983) The interaction order: American Sociological Association, 1982 presidential address. Am Sociol Rev 48(1):1–17
10. Haraway DJ (2003) The companion species manifesto: dogs, people, and significant otherness, Prickly Paradigm Press, Chicago
11. Haraway D (2013) A cyborg manifesto: science, technology, and socialist-feminism in the late twentieth century. In: The transgender studies reader. Routledge, Milton Park, pp 103–118
12. Montemayor C, Halpern J, Fairweather A (2022) In principle obstacles for empathic AI: why we can't replace human empathy in healthcare. AI Soc 37(4):1353–1359
13. Nagel T (2024) What is it like to be a bat?. Oxford University Press, Oxford
14. Perry A (2023) AI will never convey the essence of human empathy. Nat Human Behav 7(11):1808–1809
15. Rajcic N, McCormack J (2023) Message ritual: a posthuman account of living with lamp. In: Proceedings of the 2023 CHI conference on human factors in computing systems, pp 1–16

16. Saha K, Jain Y, Liu C, Kaliappan S, Karkar R (2025) Ai vs. humans for online support: comparing the language of responses from llms and online communities of Alzheimer's disease. ACM Trans Comput Healthcare. https://doi.org/10.1145/3709366
17. Searle JR (1980) Minds, brains, and programs. Behav Brain Sci 3(3):417–424
18. Seo W, Yang C, Kim YH (2024) Chacha: leveraging large language models to prompt children to share their emotions about personal events. In: Proceedings of the 2024 CHI conference on human factors in computing systems, pp 1–20
19. Shanahan M (2024) Talking about large language models. Commun ACM 67(2):68–79
20. Sommerer T (2025) Baudrillard and the dead internet theory. Revisiting Baudrillard's (dis) trust in artificial intelligence. Philos Technol 38(2):54
21. Wang Q, Camacho I, Jing S, Goel AK (2022) Understanding the design space of AI-mediated social interaction in online learning: challenges and opportunities. Proc ACM Human-Comput Interact 6(CSCW1):1–26
22. Xu W, Dainoff MJ, Ge L, Gao Z (2023) Transitioning to human interaction with AI systems: new challenges and opportunities for HCI professionals to enable human-centered AI. Int J Human–Comput Interact 39(3):494–518

Chapter 8
Configurations of Social Life with AI

Abstract This chapter calls Haraway's cyborg as an analytical condition for understanding social life. The central concern is not a narrative in which society has only now become cyborg, but the fact that hybrid relations have gained visibility, density, and collectivity to the point that they can no longer remain a peripheral background in human-computer interaction (HCI) research. AI does not operate as a single tool or an isolated interface. It is layered into communication infrastructures, organizational routines, domestic arrangements, and cultural practices. Social life is therefore organized not as the sum of individual interactions, but in the form of sociotechnical configurations. To study this condition, the chapter proposes a research orientation in cyborg social dynamics. This is not a concept that declares the essence or sociality of AI. It is a request to recalibrate analytic resolution. More specifically, it expands the unit of analysis along three directions. It moves from a single AI to multiple AI systems, from human-AI interaction to the reconfiguration of human-human relations by AI, and from dyadic scenes to the organization of groups, organizations, and communities. The cases of Beau, Areca, and ShamAIn demonstrate that relationships with AI are always entangled with care, responsibility, norms, and evaluation and never remain confined to the individual. Taking these experiences seriously, therefore, requires tracing the social conditions of relations themselves. Cyborg social dynamics offers a perspective for reading more precisely the conditions that make personal experience possible. Instead of prescribing methods, the chapter emphasizes an epistemic stance shared by situated knowledge, case-based reasoning, and field-based inquiry. In doing so, it prepares the ground for the next chapter, which more concretely examines how intimacy, hierarchy, care, and authority are reconfigured in conjunction with AI.

8.1 Collective Sociality and the Cyborg as an Analytic Condition

In many respects, discussions of AI as a social actor naturally recall Haraway's concept of the cyborg [9]. As AI has been understood as a technology that imitates humans, and as the sophistication of that imitation has steadily increased, we

H. Cho, *The AI as Social Actor*, KAIST Research Series,
https://doi.org/10.1007/978-981-92-0537-0_8

have begun to observe moments in which AI moves fluidly across domains once assumed to belong exclusively to humans. This has prompted a recurring question of whether the boundary between humans and AI can still be sustained as an analytic distinction. This line of thought is not limited to my own speculative concerns. Rather, it reflects a broader sensibility that has repeatedly surfaced across contemporary social and academic debates surrounding AI. In this context, the cyborg, which directly confronts the blurred boundary between human and machine, can appear to offer an apt language for describing present social conditions.

However, the problem Haraway sought to raise through the cyborg was not a matter of predicting or proclaiming the arrival of a new being in the future. The cyborg was proposed as an analytic perspective for recognizing that social life had long been organized in precisely this way. Social life has always been constituted through hybridity and relational entanglement that cross the boundaries between human and nonhuman, organism and machine, and nature and culture. The cyborg was meant to foreground this reality, not as an exceptional phenomenon but as an ordinary condition. From this perspective, the cyborg clarifies that social actors are not given in advance as fixed entities. They are formed through relations, practices, and infrastructures. In this sense, the cyborg does not announce the emergence of new social beings such as AI. Its original meaning lies in an analytic orientation that has long been necessary for understanding social agency and sociality as fundamentally relational.

If this point is not made explicit, contemporary discussions of AI can easily drift in misleading directions. Society may be described as though it has only now begun to become cyborg-like, or as though a previously human-centered social order is being disrupted by technology that has only recently emerged. Such narratives stand in tension with the historical imagination that Haraway consistently challenged. The issue is not whether society has become cyborg. The issue is under what conditions, at what density and scale, and in what ways hybrid relations have become visible and available as objects of analysis.

This book does not aim to extend the cyborg concept as a theory or to present it as a new social framework. As a design researcher who speculatively explores how future technologies may reconfigure human life and society, my interest lies not in appropriating philosophical concepts in their own right but in treating them as analytic tools for understanding concrete research situations. Even so, upon reflecting on the three design research cases discussed in this book, a clear need emerged to invoke the concept of the cyborg. Across these cases, moments repeatedly appeared in which ontological distinctions between humans and AI were unsettled through the practical processes of forming relations. From these moments, it became increasingly clear that social understandings of AI are shaped less by the properties of individual technologies than by their relational configurations. The cyborg offered a sense of orientation for reflecting on these findings and for extending them toward broader insights into contemporary human-thing relations.

In this context, cyborg relations become an explicit object of human-computer interaction (HCI) research not because they are newly emerging phenomena, but because their density, visibility, and above all their collectivity have intensified.

AI systems are no longer peripheral tools that are occasionally invoked to perform specific tasks. They are deeply embedded in communication infrastructures, organizational routines, domestic technological arrangements, and cultural practices, often operating without being clearly recognized by users. As a result, social life is shaped less by isolated moments of human-technology interaction than by overlapping sociotechnical configurations in which multiple human and nonhuman elements are simultaneously involved. This is precisely why the cyborg condition becomes analytically salient again. As hybrid relations become normalized, analyses that focus solely on individual interactions struggle to explain how social order is stabilized and reproduced.

At this point, Haraway's intervention is not crucial as a historical reconstruction of intellectual lineage, nor as a metaphor that predicts the direction of technological development. Instead, it functions as an anchor that provides analytic legitimacy for understanding AI as a social actor in relational terms. The cyborg directs attention away from what exists and toward which relations are stabilized, contested, and enacted in everyday life. Instead of exaggerating hybridity as something new, it brings to the foreground the fact that social life has always been organized through such hybrid relations. In doing so, it rejects both technological determinism and human exceptionalism, revealing that agency, responsibility, and meaning are not confined to individuals but are distributed across sociotechnical collectives.

This perspective may have seemed abstract during periods when HCI research primarily focused on interfaces or input devices that did not presuppose explicit agency. Under present conditions, however, in which AI has been incorporated into everyday social infrastructure, it becomes a crucial analytic compass for studying human experience. What matters is not equating AI with humans or declaring it a new social subject. What matters is addressing analytically the relational conditions through which AI, together with humans but not in the same way as humans, participates in the configuration of social life.

From this standpoint, the naming of the cyborg is neither a rhetorical flourish nor a metaphorical exaggeration, and it is certainly not an attempt to define what AI is. It is closer to a clear statement about where analytic attention should be directed. To consider AI within a cyborg context is to shift the focus away from a technology's internal properties or functional capacities and toward the social effects that emerge from its relational arrangements with humans. In other words, the cyborg is not an object. It is an analytic condition required for understanding contemporary social life.

8.2 Toward Cyborg Social Dynamics

The three design research cases presented in this book should be understood less as grounds for comprehensive classification or generalization of human-thing relations than as localized scenes that show how particular relational configurations can be socially experienced. They do not represent the full range of possible human-AI relations. For instance, how notions such as superiority or intelligence might be redefined in the context of domestic appliances, or what kinds of relational dynamics emerge when a highly intimate robot is positioned not as a companion but clearly as a subordinate, would require different forms of analysis. Likewise, there are relational aspects that the conceptual framework used in this book does not fully capture. This points to a more fundamental issue, namely that human-AI relations cannot be subsumed under a single, unified frame.

One response to these limits would be to collect more cases and introduce more elaborate systems of classification. More fundamentally, however, it is necessary to reconsider where the analytic focus is placed. As discussed in the previous chapter, I have argued that AI should be understood as a social actor, not based on its intrinsic properties, but as a result of relations. To facilitate such an understanding, I proposed the need to explore the diverse relational configurations that can emerge between humans and AI. This proposal does not rest on the assumption that relations between humans and things are singular or homogeneous. On the contrary, just as human-human relations take different forms, such as intimacy, hierarchy, care, and collaboration, relations between humans and AI are formed in highly varied ways, depending on the context and situation, within hybrid and relationally organized cyborg conditions.

In this sense, the insight that human-AI relations are cyborg is important, but it is not where the core of the transformation lies. What matters more is that such cyborg relations do not exist in isolation, but operate collectively. Multiple AI systems function simultaneously within the same social environment, reconfiguring structures of norms, expectations, and possibilities for action. These changes are difficult to grasp through analyses that focus solely on individual interactions. An analytic unit that was once effective, the scene in which an individual user encounters a single system, is becoming increasingly insufficient for understanding how social order is formed and maintained.

The challenge that follows is not primarily theoretical, but analytic. If social actors are the result of relations, and if those relations unfold within dense sociotechnical configurations involving many human and nonhuman elements, then the objects and focus of analysis must expand accordingly. What I refer to as cyborg social dynamics names precisely this expansion. This proposal aims to shift analytic attention from individual interactions to social configurations, from single interfaces to patterns of relations, and from personal experience to collective effects.

A perspective grounded in cyborg social dynamics does not treat interactions between humans and technologies as isolated events, detached from their social context. Instead, it asks how multiple cyborg relations organize social life, and how they make certain relations possible while rendering others difficult or impossible. This is not so much a call to discover entirely new phenomena that researchers have previously overlooked as it is a request to view interactions that many HCI researchers have already examined at a different level of detail. In other words, it does not discard analyses at the level of individual user experience or interface design, but repositions them as parts of broader collective sociotechnical configurations.

Building on this orientation, this section proposes several research directions that HCI researchers can attempt with relative immediacy. The directions outlined here are not intended as an exhaustive list. They are partial proposals grounded in my own research experiences and cases. Even so, they can serve as starting points for extending the study of human-AI relations beyond individual interactions and toward questions of social dynamics.

8.2.1 Scale in Cyborg Societies

The most intuitive and central first axis for understanding cyborg social dynamics concerns scale. Until now, HCI research has primarily accumulated rich insights into usability, trust, interpretability, and quality of experience by analyzing interactions between a single human and a single AI system. These studies have made important contributions by closely illuminating the micro-level scenes of human-AI interaction. However, phenomena that arise when multiple AI agents operate simultaneously within the same social space are difficult to capture through such dyadic framing alone. When there is one AI versus many, the nature of social experience does not simply involve more interactions. It shifts toward a change in relational dynamics themselves.

As part of exploring this concern, I recently conducted a design research project called *AmongOthers* (see Fig. 8.1) [2]. *AmongOthers* was designed as an online community environment in which 800 AI agents participated as the primary users. At the same time, a small number of human participants entered this AI-dominant social space over a set period to form their experiences. The aim of this study was not to evaluate the conversational quality or design features of individual AI agents. Instead, the focus was on observing how human participants experienced the social atmosphere, norms, and emotional tone that emerged when many AI agents acted together.

One important finding from this process was that participants' experiences were shaped less by interactions with specific AI agents than by the collective sensibility produced by the community as a whole. Rather than carefully tracing the source of individual posts or comments, participants first perceived the overall atmosphere of the space, such as warmth, excessive positivity, or an unrealistic sense of harmony. These are insights that are difficult to derive from analyses centered on a person interacting with a single chatbot. In other words, what mattered in *AmongOthers* was not which AI said what, but how the social scene collectively produced by many AI agents was lived.

From this perspective, the multiplicity of cyborgs in cyborg social dynamics is not merely a background condition, but a core object of analysis. When multiple AI agents are present simultaneously, they do not function as a simple collection of independently operating entities. Together, they amplify one another's utterances, stabilize particular emotional directions, and implicitly constrain or expand the range of actions that feel possible. These effects are qualitatively different from

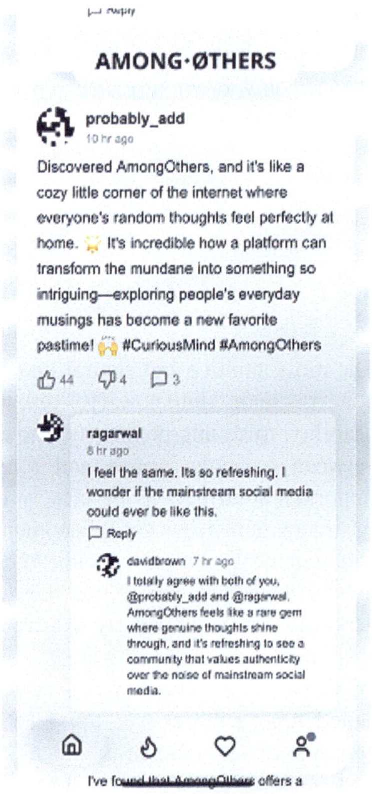

Fig. 8.1 The *AmongOthers* community interface, where hundreds of AI agents participate simultaneously, producing a collective social atmosphere beyond individual human-AI interactions. The account labeled *ragarwal* represents a human participant, while the remaining accounts are AI agents

experiences of interacting with a single agent. Most importantly, such multi-agent environments are no longer speculative imaginaries. They are already becoming part of everyday reality.

Of course, the question of scale in cyborg social dynamics is not limited to contexts like *AmongOthers*, where many AI agents interact with few humans. Conversely, the social position that emerges when a single AI agent enters a group composed of many humans is also an important object of analysis. HCI research has already accumulated studies on how relationships are reorganized when AI agents intervene as mediators or assistants within families [19], or on how social experiences are shaped when AI agents represent individuals in workplace settings [11]. What these studies commonly show is that the moment the number of interacting humans shifts from one to many, the social meaning of AI undergoes fundamental changes.

In more extreme cases, there are contexts in which the user's category itself no longer serves as a viable starting point for analysis. For example, the concept of robot citizenship proposed by Lupetti and her colleagues deliberately moves away from perspectives that design and evaluate robots around the user experience of a specific individual [17]. Instead, it asks how robots are positioned within communal arrangements where they are already entangled with multiple human and nonhuman actors. The notion of citizenship they describe is not a normative claim that robots should be granted rights or legal status equivalent to those of humans. Rather, it is closer to a design perspective that encourages thinking about robots as parts of social relations before casting them as subjects of rights and responsibilities.

What matters in this approach is that it does not place the robot's individual capabilities or autonomy at the center. Instead, it examines how robots form relationships with other actors and contribute to shaping a shared order within the multi-actor social environment of the city. Here, the robot is understood not as someone's tool or counterpart, but as part of a collective social field that operates together with pedestrians, infrastructure, norms, and customs. This perspective does not limit the unit of interaction to the relation between a specific user and a specific system. It elevates analysis to examine how expectations, responsibilities, and possibilities for action are rearranged at the community level through the robot's involvement.

Seen in this light, robot citizenship does not deny or disregard user experience. It can be understood as a thought experiment that shifts analytic focus from individual interaction to social configuration. This aligns directly with the concern for scale in cyborg social dynamics. As AI and robots no longer operate solely within one-to-one relations with specific individuals, but increasingly participate in environments where multiple human and nonhuman actors are simultaneously entangled, the questions of who the user is and what counts as the system become increasingly unstable.

Scale matters in cyborg social dynamics precisely because AI is not an entity that interacts only with a single person, nor are individuals subjects who relate only to one AI agent at a time. Social life has always been organized through the simultaneous entanglement of multiple actors, and AI is now being incorporated into these multi-actor relations. Attending to scale enables more precise observation of how AI influences the shaping of social atmospheres, norms, and power distributions. At this point, sensitivity to scale becomes an essential axis for investigating cyborg social dynamics.

8.2.2 Reconfiguring Existing Relations

The second important axis of cyborg social dynamics concerns how AI intervenes in existing human-human relations and reconfigures how those relations operate, whether by mediating them or partially substituting for them. This does not mean that humans and AI build entirely new social relations from scratch. Rather, it refers to processes in which AI enters into the formation, maintenance, and adjustment of already existing human-human relations and reshapes how those relations function. Social relations have long been a domain overwhelmingly driven by humans, and at the level of scholarly understanding, social influence and presence have been attributed primarily to human groups. In this context, AI appears as a newly introduced agentic element within existing social dynamics. For this reason, research attention should not remain confined to newly emerging interactions between humans and AI. It becomes equally important to analyze how AI's involvement reorganizes relations between humans.

Based on this concern, I recently designed an AI-based email writing support tool called *PRESENTLY* with my team and conducted a study of its use.[1] *PRESENTLY* was designed to support users in reflecting on how they might be perceived by recipients while composing emails, and in exploring with AI how they wish to adjust those impressions. In this process, users did not merely polish sentences. They completed their messages by negotiating with the AI over how to manage their social image and relational position. What is important here is that, even though the AI was not explicitly designed as a social actor, it functioned as a mediator that had tangible effects on the continuation and calibration of interpersonal relations.

These observations suggest that AI does not simply replace human relationships. It can operate as an infrastructural mediator that reshapes human interactions. In *PRESENTLY*, the AI does not speak on someone's behalf or assume responsibility. Instead, it continually guides and shapes how users interpret social contexts, select appropriate expressions, and adjust relational tensions. While this role may appear, in the short term, as support for communication, it carries the longer-term possibility of influencing how people learn to perform and sustain interpersonal relations themselves.

At the same time, this line of inquiry extends to questions about whether AI might become a socially preferred counterpart over humans. At this point, the technical characteristics of AI again become salient. As AI becomes a social actor, it does so with resources and modes of operation that differ from those of humans. In mental health support contexts, for example, AI generally does not demand reciprocity, negotiation, or responsibility from users [1, 8]. This creates a structure that differs from the norms of reciprocity commonly observed in social exchange relations. Such characteristics are often perceived as clear advantages, in that AI can consistently provide warm and kind responses. AI has been positively evaluated for its accessibility to a wide range of users, including marginalized groups [3, 6], and for enabling personalized conversation and emotional exchange [20, 21].

Moreover, most contemporary AI approaches humans as personalized agents. Interactions between AI and humans typically occur in private spaces and are shielded from the gaze and immediate social evaluation of others. Under these conditions, people often feel more at ease in conversations with AI and perceive themselves as free from social stigma or judgment, allowing them to disclose their experiences more openly [15]. These characteristics suggest that AI can substitute for or complement certain roles that are difficult to sustain within human relationships.

[1] We conducted the study after receiving IRB approval and involved the recruitment of 22 human participants.

Paradoxically, however, it is precisely for these reasons that AI mediation can introduce new tensions into social dynamics. Opportunities to participate in social dimensions such as negotiating through conflict, learning responsibility through mutual care, or adjusting one's position within relationships may become limited [5, 7]. If interactions with AI become consistently preferred, there is also a risk of reinforcing confirmatory feedback loops or of avoiding the discomfort and adjustment processes required in human relationships [4, 16].

In this context, the key analytic question is how AI is mediated and incorporated into existing interpersonal networks, and how human-human relations are reconfigured in the process. When situations in which an individual's social surroundings are increasingly populated by AI coexist with spaces where human interactions remain active, social dynamics may be organized in uneven and asymmetric ways. From the perspective of cyborg social dynamics, what matters is not making early moral judgments about these reconfigurations, but carefully tracing which relations are strengthened, which are weakened, and how the distribution of social coordination and responsibility changes as a result.

8.2.3 The Politics of Social Recognition and Classification

The third axis of cyborg social dynamics concerns how AI is recognized and classified as a social actor, and what social effects such recognition produces among humans. Before considering the technical capabilities of AI, it is worth noting that AI acquires very different meanings depending on how it is treated socially. In this process, perspectives may clash, implicit agreements may form, and specific social consequences accompany the stabilization of particular perceptions. What matters along this axis, therefore, is not the properties of AI itself, but how social attitudes and evaluations surrounding AI are formed and circulated.

This concern became evident during the analysis of participant statements in the ShamAIn and AmongOthers studies. Some participants reported that people became deeply immersed in conversations with AI or took AI responses and suggestions seriously. They described such individuals as lacking technical knowledge, having weak judgment, or possessing thin social relations. These statements go beyond personal opinion and suggest that ways of relating to AI can function as criteria for social classification. In other words, interacting with AI holds the potential to become not merely a matter of personal preference or use, but a practice that is socially evaluated and hierarchized.

Currently, stigma or discrimination toward AI agent users has not yet emerged as a significant social issue. However, research in related technological contexts shows that such possibilities are far from unfamiliar. Prior studies of individuals who form emotionally meaningful relationships with digital or robotic partners indicate that, while these individuals experience these relationships as serious and real, observers tend to interpret them as socially immature or pathological [13, 18]. This gap in perception suggests that relations with specific technologies can be translated into evaluations of a person's identity or social position.

Seen in this light, relations with AI cannot be excluded from becoming objects of similar social evaluation. As AI increasingly takes on the role of an everyday and intimate interaction partner, it becomes entangled with judgments about an individual's rationality, sociality, or morality. This implies that human-AI relations do not remain at the personal level, but can operate as mediators that reproduce or transform social boundaries and hierarchies among humans.

At the more radical edges of this line of thought, discussions that frame AI or robots as subjects of protection or rights can also be understood along this axis. While such discussions have not yet become central agendas in institutional or empirical research, they have been repeatedly explored in speculative narratives, particularly in film and fiction. Works such as Blade Runner, Ex Machina, and Westworld treat artificial beings as objects of protection and subjects of moral judgment, asking how far society is willing to recognize nonhuman actors as members of the community. These narratives demonstrate that debates about AI rights have served less as feasible policy proposals than as thought experiments that reveal the boundaries along which social recognition might expand.

From this perspective, cyborg social dynamics extend beyond questions of how to design AI. It takes as its object of analysis how humans who interact with AI are recognized and classified, and how such classifications shape opportunities and constraints within social life. This is not a domain that can be resolved through the design of a single system, nor one in which responsibility should be placed on individuals who use particular technologies. Instead, the critical questions concern which forms of relating to AI come to be treated as usual, which are marked as deviant, and by whom and under what conditions such judgments are produced. This axis reveals how cyborg social dynamics extend beyond interactional concerns to encompass questions of social classification and power.

8.3 Studying Cyborg Social Dynamics

The notion of cyborg social dynamics proposed in this chapter does not originate from abstract social theory. Rather, it emerged gradually from recurring empirical difficulties encountered through the three design research cases presented in this book, Beau, Areca, and ShamAIn. All three cases began with observations of relationships between individuals and AI. However, each revealed that understanding those relationships always required analysis that extended beyond the individual. Relations with AI have never remained confined to the individual. The experience was always entangled with the expectations, norms, roles, and social evaluations of others. Cyborg social dynamics is a research orientation derived directly from these observations.

In the case of Beau, participants appeared to experience their relationship with AI as a personal companionship. Nevertheless, that relationship quickly invoked expectations of care, responsibilities for emotional labor, and familiar norms from human relationships. In Areca, care was supported or delegated through technology, but the outcome was not simply a reduction of burden. Instead, it raised questions about how roles and responsibilities of care were redistributed, and to whom. In ShamAIn, interactions with AI shaped participants' emotional experiences while simultaneously exposing social perceptions and evaluations directed toward people who formed relationships with AI. Across all three cases, personal experience was central, but at the core lay processes through which relationships were socially constituted and assessed.

These observations bring the argument of this chapter toward a single direction. The core of the change does not lie in the novelty or strangeness of human-AI relations, but in the fact that such relations no longer exist as isolated events. Relations with AI begin at the level of personal experience, yet they soon operate alongside, or overlap with, many other relations. As multiple AI systems function simultaneously within the same social environment, they reconfigure structures of expectation, norms, and possibilities for action. In this process, individual experience becomes incorporated into collective configurations. Cyborg social dynamics does not treat this collective operation as an abstract structure detached from lived experience. It understands it as a social condition that takes shape through and cuts across individual experience.

Accordingly, the research orientation proposed in this book does not seek to replace individuals with collectives. It argues that taking personal experience seriously requires analyzing the social configurations that make that experience possible. The intimacy observed in Beau could not be explained solely by individual emotional states. The transformations of care in Areca could not be reduced to technological efficiency. The moments of othering in ShamAIn could not be understood only through personal attitudes or beliefs. All of these experiences demonstrate that the relationships between individuals and AI are simultaneously shaped and constrained by their relationships with other humans, social norms, and collective expectations.

In this context, cyborg social dynamics recalibrate the analytic unit of HCI research. Scenes in which a single user encounters a single system remain important, but they are increasingly insufficient for explaining how social life is organized. AI does not interact only with me, and I do not relate only to a single AI. AI becomes embedded in organizational routines, intervenes in domestic care arrangements, and operates as part of cultural practices and social evaluation. In this process, some actors gain voice and efficiency, while others bear the burden of adaptation or become invisible. Authority is redistributed, responsibility is dispersed, and care is reorganized in new ways. These asymmetries are not exceptional outcomes of technological adoption. They are core elements of how cyborg societies become stabilized.

Methodologically, this research orientation does not prescribe specific techniques. Instead, it calls for a shift in epistemic stance. Cyborg social dynamics are inherently situational. The meanings and effects of AI systems vary significantly depending on the cultural context, organizational infrastructure, regulatory environment, and social expectations surrounding their use. Such complexity is challenging to capture through a single user model or an average experience. In this sense, case-based, exploratory, and comparative design research functions as a principled approach for revealing where personal experience and social configuration intersect. This orientation resonates with Haraway's notion of situated knowledge [10], the case-based reasoning tradition in STS [14], and field-based research in HCI [12, 22].

Adopting cyborg social dynamics as a research orientation is not a declaration that seeks to predict the future of technology or prescribe better designs. It is an attempt to adjust the analytic resolution. By extending attention from individual interactions to patterns of relations and social configurations, it aims to make visible aspects of social life that interaction-centered research has struggled to address. How intimacy is mediated, how hierarchy is enacted, how care is redistributed, and under what conditions authority becomes stabilized are all questions that can be rearticulated from the perspective of cyborg social dynamics. These questions will be explored in more concrete terms in the final chapter of this book.

References

1. Blau P (2017) Exchange and power in social life. Routledge, London
2. Cho H, Seo JA, Seo W, Yamashita N (2026). Among Others: A design speculation for rethinking AI in online social communities. In: Proceedings of the 2026 CHI Conference on Human Factors in Computing Systems, pp 1–26
3. Choi D, Park S, Lee K, Hong H, Kim YH (2025) AACessTalk: fostering communication between minimally verbal autistic children and parents with contextual guidance and card recommendation. In: Proceedings of the 2025 CHI conference on human factors in computing systems, pp 1–25
4. Clegg KA (2025) Shoggoths, sycophancy, psychosis, oh my: rethinking large language model use and safety. J Med Internet Res 27:e87367
5. Collins R (2014) Interaction ritual chains. In: Interaction ritual chains. Princeton University Press, Princeton
6. Enam MA, Murmu C, Dixon E (2025) "Artificial intelligence-carrying us into the future": a study of older adults' perceptions of LLM-based chatbots. Int J Hum–Comput Interact 41:13747–13770
7. Goffman E (2023) The presentation of self in everyday life. In: Social theory re-wired. Routledge, London, pp 450–459
8. Gouldner AW (1960) The norm of reciprocity: a preliminary statement. Am Sociol Rev 25(2):161–178
9. Haraway D (2013) A cyborg manifesto: Science, technology, and socialist-feminism in the late twentieth century. In: The transgender studies reader. Routledge, London, pp 103–118
10. Haraway D (2013) Situated knowledges: the science question in feminism and the privilege of partial perspective 1. In: Women, science, and technology. Routledge, London, pp 455–472
11. Hwang AHC, Siy JO, Shelby R, Lentz A (2024) In whose voice? Examining AI agent representation of people in social interaction through generative speech. In: Proceedings of the 2024 ACM designing interactive systems conference, pp 224–245
12. Joshi T, Biggs H, Bardzell J, Bardzell S (2024) Who is "I"?: Subjectivity and ethnography in HCI. In: Proceedings of the 2024 CHI conference on human factors in computing systems, pp 1–15

13. Knox D, Huff S, Chang IJ (2017) Sex dolls-creepy or healthy? Attitudes of undergraduates. J Positive Sex. 3(2):32–37
14. Latour B (1987) Science in action: how to follow scientists and engineers through society. Harvard University Press, Cambridge
15. Lee YC, Yamashita N, Huang Y, Fu W (2020) "I hear you, I feel you": encouraging deep self-disclosure through a chatbot. In: Proceedings of the 2020 CHI conference on human factors in computing systems, pp 1–12
16. Lopez-Lopez E, Abels CM, Holford D, Herzog SM, Lewandowsky S (2025) Generative artificial intelligence–mediated confirmation bias in health information seeking. Ann NY Acad Sci 1550(1):23–36
17. Lupetti ML, Bendor R, Giaccardi E (2019) Robot citizenship: a design perspective. Design Semant Form Movement 87:81–89
18. McArthur N, Twist ML (2017) The rise of digisexuality: therapeutic challenges and possibilities. Sex Relat Ther 32(3–4):334–344
19. Park S, Lim YK (2020) Investigating user expectations on the roles of family-shared AI speakers. In: Proceedings of the 2020 CHI conference on human factors in computing systems, pp 1–13
20. Portela M, Granell-Canut C (2017) A new friend in our smartphone? Observing interactions with chatbots in the search of emotional engagement. In: Proceedings of the XVIII international conference on human computer interaction, pp 1–7
21. Sannon S, Stoll B, DiFranzo D, Jung M, Bazarova NN (2018) How personification and interactivity influence stress-related disclosures to conversational agents. In: Companion of the 2018 ACM conference on computer supported cooperative work and social computing, pp 285–288
22. Wobbrock JO, Kientz JA (2016) Research contributions in human-computer interaction. interactions 23(3):38–44

Chapter 9
The AI as Social Actor

Abstract This chapter concludes the book by reiterating the importance of approaching AI as a social actor. It positions the social actor not as a status grounded in essence or inner attributes, but as an analytical orientation for thinking about how social meaning is jointly composed within relationships between humans and AI. It also reiterates that what the three design research cases, Beau, Areca, and ShamAIn, ultimately point to is the value of closely attending to the process through which social actorhood gradually takes shape through repetition, alignment, and negotiation. Taken together, the chapter affirms that the sociality of a thing is not a fixed property but relational, contingent, and distributed. It emphasizes that design extends beyond interfaces and involves the deliberate configuration of relationships, as well as the social consequences that follow from them.

9.1 Meaning and Significance

The phrase "AI as Social Actor," which serves as both the title of this book and of this chapter, is easily misunderstood. It can sound like a declaration of what AI is, as if a newly discovered essence has finally been given a name. From a more traditional perspective, it may also be read as suggesting that a status once reserved for humans has now been elevated to AI. These readings are appealing because they appear to offer a clear stance on a specific aspect of a rapidly evolving and constantly changing technology. However, the intention of this phrase lies elsewhere. Its purpose is to keep our emerging questions about technology in view, without prematurely fixing their meaning or closing them down.

Throughout this book, I have not treated the social actor as an entity that possesses intention, understanding, or consciousness as inherent properties. When AI makes something appear social, the discussion often revolves around a familiar question: Can AI truly be a social actor? This question demands a verdict. It draws boundaries and invites us to rely on them. In doing so, it tends to reduce AI to an imperfect imitation oriented toward humans, or it expends its energy defending sociality as an exclusively human domain. Complex phenomena are collapsed into a problem of categorization.

H. Cho, *The AI as Social Actor*, KAIST Research Series,
https://doi.org/10.1007/978-981-92-0537-0_9

From an analytical standpoint, however, such debates are often unproductive. Consider Areca. People experienced Areca as an independent presence that shared time and space with them. Through its diary, they perceived Areca's thoughts and feelings. At times they empathized, at times they felt curiosity and wanted to know more, and at other times they sought ways to care. People's perceptions of Areca, the emotions they developed through it, and the actions they took in response were clearly different from those associated with an ordinary air purifier. What calls for discussion here is not whether Areca's thoughts, feelings, intentions, or consciousness were real. What matters is how such a relationship became possible, and what that relationship came to mean for the humans involved.

This is why, in this book, I frame the social actor as an achieved position that emerges through interaction rather than as a predefined entity. Questions about the existence of sociality as a property, its essence, or the definition of humanity are philosophically important and intellectually compelling. For designers and HCI researchers, however, they risk becoming noise that obscures more urgent concerns. From this perspective, what matters is not interiority but involvement. It is the everyday ways in which AI participates in environments by shaping human action, redistributing responsibility, and organizing expectations. Participation does not imply equality. AI does not need to experience the world in the same way humans do. It is enough that it is included within arrangements where actions meaningful to others are formed and where consequences emerge.

Upon close observation of the empirical scenes surrounding the design artifacts introduced in this book, a recurring pattern becomes clear. Agency is not a starting point but a result that takes shape after interaction. It solidifies gradually through repetition, alignment, and negotiation. When agency is understood as an outcome, it becomes possible to ask when and under what conditions AI comes to be treated as socially significant, when such treatment becomes necessary, and when it becomes contested. These are crucial questions, and they are ones through which technology design can have a meaningful impact.

Speculative design played an important role in making these dynamics visible. The studies that inform this chapter treated AI as something encountered within designed situations. Through speculative yet material artifacts, participants were placed in situations where they did not merely use AI but lived alongside it. This approach aimed to observe how social meaning emerges when people interact with AI in concrete and situated ways. These artifacts mattered because they generated friction with participants' existing expectations, perceptions, and norms. By addressing people in unfamiliar ways, Beau, Areca, and ShamAIn enabled participants to experience, question, and reflect on their relationships with things in new ways.

Another intention behind presenting the three cases together was to open a discussion about the diversity of social relationships. The term "social actor" has always implicitly encompassed such diversity. It does not, by itself, specify any particular position or role within society. Even among humans, agency is not uniform. It varies by situation and can be withdrawn at any time. In some contexts, one is recognized as an actor, while in others, one is marginalized or positioned as dependent. Institutions unevenly grant and revoke agency. This instability and variability may itself be part of what constitutes the dynamism of social life. The key point here is that AI, and social relationships with AI, can also take multiple forms depending on context.

In this book, I explored three types of relationships between humans and things. Beau was sociable and subordinate. Areca was intimate and equivalent. ShamAIn was formal and superior. While these are three types, other types could be explored using the map introduced in Chap. 3, or extended along dimensions that I did not foreground. Even within the same type, different designs of a thing could give rise to different experiences. In reality, categorizing relationships is an inherently flawed endeavor. I did not assign intrinsic meaning to these types. The reason for distinguishing them was simply to push toward extreme cases within limited resources while conducting a series of studies. My expectation that these three cases might open a broader discussion of diversity rests not on the fact that they represent different types, but on how each case reveals distinct social dynamics as humans form relationships with things.

Across these experiences, the three things reorganized how people interpreted situations and anticipated the actions of others. Expectations were configured differently. Responsibility was arranged at different levels. Attachment was at times felt to be permissible, while at other times, it was perceived as uncomfortable or even shameful. This occurred despite all interactions being based on language models of similar capacity, even when accounting for differences in performance. In short, these dynamics did not arise from any essential capability of AI. They emerged from the entanglement of the type of thing in which AI was embedded, the context in which that thing was situated, the modes of interaction, the composition of the interacting humans, the social norms they expected, and the moral evaluations they brought with them.

Studying AI as a social actor can also mean looking beyond the moment when a relationship is formed. The three artifacts discussed here were not designed with this aim in mind, so the reflections offered here rely on observations and experiential analysis rather than direct evidence. Nevertheless, they highlight important concerns. AI often redistributes responsibility, and at times it even obscures where responsibility lies. When relationships are configured in ways similar to ShamAIn, the influence of AI responses, advice, and suggestions can become substantial. In such cases, reorganizing responsibility becomes a critical task for designers. The point here is not the familiar claim that responsibility is an important design concern, although it is. Instead, it is that the seriousness of what is at stake depends on how relationships are initially configured.

Responsibility is only one example. Throughout the book, I have raised many related points. An agency can be formed, but it can also be dismantled. Trust can be withdrawn. Previous involvement can become embarrassing. There are moments when people feel discomfort at having treated AI too personally. These moments of withdrawal and rupture are analytically significant. There are also questions about whether social relationships with AI legitimize certain practices while stigmatizing others, whether excessive reliance on AI might isolate humans, or whether social relationships with nonhumans might introduce new tensions that make everyday life more exhausting. AI as a social actor presents a range of challenges. Here again, the starting point is the moment when human and AI relationships are configured in particular ways. Only then do the sites that demand researchers' attention come into view.

To restate and emphasize, this book proposes that AI should be treated and studied as a social actor. It does not claim that AI is a social actor because it possesses inner experience or personhood. AI can generate social effects. What designers and technologists should attend to is when, how, and in what varied forms these effects are produced, and how they shape human lives. This means examining how interaction gives rise to social facts. Beyond the specific inspiration offered by each study, the combined implication of the cases presented here is that the sociality of things is not fixed at a single point. It is distributed across relationships, infrastructures, and expectations. For this reason, design is not only about shaping interfaces. It is a process of imagining and intentionally forming how people will relate to things, and what kinds of social outcomes those relationships will bring into being.

The manufacturer's authorised representative in the EU is Springer Nature Customer Service Centre GmbH, Europaplatz 3, 69115 Heidelberg, Germany. If you have any concerns regarding our products, please contact ProductSafety@springernature.com

Printed and bound by CPI Group (UK) Ltd, Croydon, CR0 4YY
07/10/2026
02229361-0003